LNA-ESD CO-DESIGN FOR FULLY INTEGRATED CMOS WIRELESS RECEIVERS

THE KLUWER INTERNATIONAL SERIES IN ENGINEERING AND COMPUTER SCIENCE

ANALOG CIRCUITS AND SIGNAL PROCESSING
Consulting Editor*: Mohammed Ismail. *Ohio State University

Related Titles:

LOW-POWER DEEP SUB-MICRON CMOS LOGIC SUB-THRESHOLD CURRENT REDUCTION
van der Meer, van Staveren, van Roermund
ISBN: 1-4020-2848-2
OPERATIONAL AMPLIFIER SPEED AND ACCURACY IMPROVEMENT
Ivanov and Filanovsky
ISBN: 1-4020-7772-6
STATIC AND DYNAMIC PERFORMANCE LIMITATIONS FOR HIGH SPEED D/A CONVERTERS
van den Bosch, Steyaert and Sansen
ISBN: 1-4020-7761-0
DESIGN AND ANALYSIS OF HIGH EFFICIENCY LINE DRIVERS FOR Xdsl
Piessens and Steyaert
ISBN: 1-4020-7727-0
LOW POWER ANALOG CMOS FOR CARDIAC PACEMAKERS
Silveira and Flandre
ISBN: 1-4020-7719-X
MIXED-SIGNAL LAYOUT GENERATION CONCEPTS
Lin, van Roermund, Leenaerts
ISBN: 1-4020-7598-7
HIGH-FREQUENCY OSCILLATOR DESIGN FOR INTEGRATED TRANSCEIVERS
Van der Tang, Kasperkovitz and van Roermund
ISBN: 1-4020-7564-2
CMOS INTEGRATION OF ANALOG CIRCUITS FOR HIGH DATA RATE TRANSMITTERS
DeRanter and Steyaert
ISBN: 1-4020-7545-6
SYSTEMATIC DESIGN OF ANALOG IP BLOCKS
Vandenbussche and Gielen
ISBN: 1-4020-7471-9
SYSTEMATIC DESIGN OF ANALOG IP BLOCKS
Cheung & Luong
ISBN: 1-4020-7466-2
LOW-VOLTAGE CMOS LOG COMPANDING ANALOG DESIGN
Serra-Graells, Rueda & Huertas
ISBN: 1-4020-7445-X
CIRCUIT DESIGN FOR WIRELESS COMMUNICATIONS
Pun, Franca & Leme
ISBN: 1-4020-7415-8
DESIGN OF LOW-PHASE CMOS FRACTIONAL-N SYNTHESIZERS
DeMuer & Steyaert
ISBN: 1-4020-7387-9
MODULAR LOW-POWER, HIGH SPEED CMOS ANALOG-TO-DIGITAL CONVERTER FOR EMBEDDED SYSTEMS
Lin, Kemna & Hosticka
ISBN: 1-4020-7380-1
DESIGN CRITERIA FOR LOW DISTORTION IN FEEDBACK OPAMP CIRCUITE
Hernes & Saether
ISBN: 1-4020-7356-9
CIRCUIT TECHNIQUES FOR LOW-VOLTAGE AND HIGH-SPEED A/D CONVERTERS
Walteri
ISBN: 1-4020-7244-9
DESIGN OF HIGH-PERFORMANCE CMOS VOLTAGE CONTROLLED OSCILLATORS
Dai and Harjani
ISBN: 1-4020-7238-4
CMOS CIRCUIT DESIGN FOR RF SENSORS
Gudnason and Bruun
ISBN: 1-4020-7127-2

LNA-ESD CO-DESIGN FOR FULLY INTEGRATED CMOS WIRELESS RECEIVERS

by

Paul Leroux

KU Leuven,
Belgium

and

Michiel Steyaert

KU Leuven,
Belgium

Springer

A C.I.P. Catalogue record for this book is available from the Library of Congress.

ISBN-10 1-4020-3190-4 (HB) Springer Dordrecht, Berlin, Heidelberg, New York
ISBN-10 1-4020-3191-2 (e-book) Springer Dordrecht, Berlin, Heidelberg, New York
ISBN-13 978-1-4020-3190-8 (HB) Springer Dordrecht, Berlin, Heidelberg, New York
ISBN-13 978-1-4020-3191-5 (e-book) Springer Dordrecht, Berlin, Heidelberg, New York

---

Published by Springer,
P.O. Box 17, 3300 AA Dordrecht, The Netherlands.

Printed on acid-free paper

Printed in the Netherlands.

Contents

Abstract

Only a few —maybe ten or fifteen— years ago, the need for telecommunication of the average citizen was limited to the 'ordinary' telephone line. The world of telecommunication, as it develops today, is characterized by a vast expansion of applications. In the recent past, the required bandwidth was limited to 4 kHz, allowing a reasonably intelligible conversation over the phone. Today, applications range from email, real-time audio and video to online gaming etc.; the required bandwidth is several orders of magnitude higher. Moreover, the user wants the freedom to access these applications anywhere and at any time. This increased mobility is the main driving force for the wireless communication market and explains the evolution of the simple cell phones, five years ago, to the portable multimedia devices, they are turning into now.

These developments naturally import an ever growing quest for increased bandwidth and mobility which can only be enabled by an equally rapid technological evolution: the functionality of the chip-sets sustaining these applications has to grow accordingly. This compression of functionality is the main driving force behind current integration research and explains the rising popularity of CMOS. As the ubiquitous digital technology, CMOS is the explicit candidate for integrating both RF front-end, analog baseband and the digital back-end on a single die.

The presented work fits well within this CMOS integration framework. The book is conceived as a tutorial on the design of CMOS low-noise amplifiers under ESD-protection constraints. It starts with an introduction on RF terminology. Concepts like quality factors, matching, noise figure, IIP3, power gain etc. are clarified. Based on a study of receiver architectures, the main LNA requirements are derived and different LNA topologies are introduced.

After a review of ESD-protection requirements in CMOS, the common-source LNA with inductive degeneration is introduced. A thorough theoretical investigation exposes the RF performance degradation induced by the classical ESD-protection. A rigorous design optimization procedure within the bounds of the ESD constraints, is described. Two alternative RF-ESD co-design procedures are proposed which are able to improve the RF performance for designs at frequencies close to and beyond 5 GHz.

These theoretical discussions are illustrated with several implementations. Two designs, described in this work, target the very demanding GPS application. The circuits operate at 1.23 GHz and 1.57 GHz respectively and achieve noise figures in the range of 1 dB. The latter amplifier was integrated within a complete CMOS GPS receiver front-end. A third amplifier, discussed in this book, is compatible with the IEEE802.11a and HIPERLAN standard and operates at 5 GHz. The circuit is fully protected against ESD exceeding the industrial requirements. The attained noise figure is 3.5 dB.

List of Symbols and Abbreviations

Symbols

Physical

k	Boltzmann's constant (1.38×10^{-23} J/K)
q	Elementary charge (1.60×10^{-19} C)

Definitions

α	Inverse of n
α_{gd}, α_{db}, etc.	Ratio between the device capacitances and C_{gs}
α_{ind}	Series resistance per inductance
β	Current gain of a bipolar transistor
γ	Excess noise factor
ϵ_{ox}	Permittivity of the gate oxide
Γ	Reflection coefficient
δ	Parameter modelling the gate noise current
Θ	Parameters modelling the mobility degradation
κ	Elmore constant of the channel
Λ	Parameter modelling the channel length modulation
μ	Mobility
μ_s	Two-port stability parameter
ξ_x	Relative contribution of C_x
$\Phi(\omega)$	Phase turn
τ	Time constant
ω_0	Operating pulsation
ω_c	Pole related to the cascode node
ω_T	Unity current gain pulsation
A_D	Area of a diode
BW	Bandwidth
c	Gate noise - drain noise correlation coefficient

C_{bp}	Capacitance of a bonding pad
C_{ESD}	Parasitic capacitance of the ESD device
C_J	Junction capacitance per unit area
C_{gs}, C_{gd}, etc.	Device capacitances
C_{ox}	Gate oxide capacitance per unit area
C_p	Parasitic parallel input capacitance
e_n	Equivalent input noise voltage
f_0	Operating frequency
F	Noise factor
F_{buf}	Noise contribution of the output buffer
F_c	Noise contribution of the cascode transistor
f_{CLK}	Clock frequency
F_d	Noise contribution of the classical drain noise current
F_{ESD}	Noise contribution of the ESD inductor
F_g	Noise contribution of the correlated part of the gate noise current
F_L	Noise contribution of the equivalent load resistance
F_{min}	Minimum noise factor, corresponding to a noise match
F_{opt}	Optimum noise factor within the matching constraint
f_{ref}	Reference frequency
f_T	Unity current gain frequency
g_{d0}	Drain-source conductance at zero V_{DS}
g_{ds}	Output conductance of a MOS transistor
g_m	Transconductance of a MOS transistor
g_{mb}	Bulk transconductance
G_T	Transducer power gain
G_u	Conductance associated with the uncorrelated part of i_n
i_c	Part of i_n that is correlated to e_n
ICP1	Input referred 1 dB compression point
I_{DS}	Drain current
IIP3	Input referred third-order intermodulation intercept power
IIV3	Input referred third-order intermodulation intercept voltage
IM3	Third-order intermodulation ratio
i_n	Equivalent input noise current
i_{nd}	Drain noise current
i_{ng}	Non-quasi static gate noise current
I_S	Diode or BJT saturation current
I_{t1}	Snapback current
I_{t2}	Second breakdown current
i_u	Part of i_n that is uncorrelated to e_n
IV3	Third-order intermodulation intercept voltage amplitude
K_n, K_p	Transconductance parameter of a NMOS, PMOS transistor
L, W	Channel length and width of a MOS transistor
L_d	Drain or load inductance

L_{eff}	Effective channel length of a MOS transistor
L_{ESD}	ESD protection inductance
L_g	Gate inductance
$L_{g,eq}$	Equivalent gate inductance
$L_{g,p}$	Equivalent parallel gate inductance
L_{min}	Minimum channel length of a MOS transistor
L_s	Source inductance
M	Miller amplification factor ($= \frac{v_c}{v_{gs}}$)
M_J	Exponent for modelling the voltage dependent junction capacitance
n	Factor modelling the bulk effect
NF	Noise figure
NF_{sys}	System noise figure
P_{av}	Available power
$P_{av,n}$	Available noise power of the source
$P_{av,s}$	Available signal power of the source
P_D	Perimeter of a diode
P_{DC}	DC power consumption
P_{in}	Input power
$P_{n,eq}$	Equivalent input referred noise power
P_{out}	Output power
Q	Quality factor of a network
Q_{in}	Input quality factor, passive input voltage amplification: $\frac{v_{gs}}{v_s}$
Q_{LNA}	Quality factor of the LNA: $\frac{f_0}{3\text{ dB BW}}$
$R_\square$	Sheet resistance
R_{bal}	Drain ballasting resistance
r_{ds}	Output resistance of a MOS transistor
R_f	Feedback resistance
R_g	Resistance of the gate fingers
$r_{g,NQS}$	Non-quasi static gate resistance
R_{in}	Input resistance
R_L	Equivalent load resistance
R_n	Resistance associated with the equivalent input noise voltage e_n
R_p	Equivalent parallel resistance of a passive element
R_S	Source resistance
$R_{S,eq}$	Equivalent source resistance, as seen by the transistor
$R_{S,opt}$	Optimal source resistance for optimal noise performance
$R_{S,p}$	Equivalent parallel source resistance
R_T	Termination resistance
S_{21}, S_{12}	Forward and reverse gain
S_{11}, S_{22}	Input and output reflection
T	Absolute temperature
t_{ox}	Thickness of the gate oxide

V_{bd}	Breakdown voltage
V_{bi}	Built-in voltage of a junction
v_{ds}, v_{gs}	Small signal drain to source and gate to source voltages
$V_{DS,sat}$	Drain to source saturation voltage
V_{GST}	DC Gate-source over-drive voltage, i.e. $V_{GS} - V_T$
v_s	Source voltage
V_T	Threshold voltage of a MOS transistor
V_{t1}	Snapback voltage
V_{t2}	Second breakdown voltage
Y_c	Correlation admittance (ratio between i_c and e_n)
Y_{opt}	Source admittance corresponding to a noise match
Y_s	Source admittance
Z_c	Characteristic impedance
Z_{in}	Input impedance
Z_S	Source impedance
$Z_{S,eq}$	Equivalent source impedance

Abbreviations

A/D	Analog to Digital
ADC	Analog to Digital Converter
AGC	Automatic Gain Control
BiCMOS	Bipolar Complementary Metal Oxide Semiconductor
BJT	Bipolar Junction Transistor
BPF	Band-Pass Filter
CDM	Charged Device Model
CDMA	Code Division Multiple Access
CG	Common-Gate
CLK	Clock signal
CMFB	Common Mode Feedback
CMOS	Complementary Metal Oxide Semiconductor
CS	Common-Source
DR	Dynamic Range
DSP	Digital Signal Processor
DSSS	Direct-Sequence Spread Spectrum
ESD	Electro-Static Discharge
FCC	Federal Communications Commission
FOM	Figure Of Merit
GCNMOS	Gate-Coupled NMOS
GGNMOS	Grounded-Gate NMOS
GPS	Global Positioning System

GSM	Global System for Mobile Communications
HBM	Human Body Model
HF	High Frequency
HPF	High-Pass Filter
IC	Integrated Circuit
IF	Intermediate Frequency
IMFDR	Intermodulation-Free Dynamic Range
IMRR	Image Rejection Ratio
I/O	Input/Output
LAN	Local Area Network
LF	Low Frequency
LNA	Low-Noise Amplifier
LO	Local Oscillator
LPF	Low-Pass Filter
MCM	Multi-Chip Module
MIM	Metal Insulator Metal
MM	Machine Model
NMOS	n-channel MOSFET
NQS	Non-Quasi Static
OFDM	Orthogonal Frequency-Division Multiplexing
OPAMP	Operational Amplifier
OSR	Oversampling Ratio
PAE	Power Added Efficiency
PCB	Printed Circuit Board
PD	Phase-Detector
PFD	Phase-Frequency Detector
PLL	Phase Locked Loop
PMOS	p-channel MOSFET
PSD	Power Spectral Density
RF	Radio Frequency
SAW	Surface Acoustic Wave
SCR	Silicon-Controlled Rectifier
SNDR	Signal-to-Noise-and-Distortion Ratio
SNR	Signal-to-Noise Ratio
SOI	Silicon On Insulator
TIA	TransImpedance Amplifier
TLP	Transmission Line Pulse
UMTS	Universal Mobile Telecommunications System
UWB	Ultra-Wide-Band
VCO	Voltage-Controlled Oscillator
VGA	Variable-Gain Amplifier
WLAN	Wireless Local Area Network

Chapter 1

Introduction

Our knowledge and understanding of the world and its nature is growing rapidly. So does our ability to control and shape it. This potential to change the world around us —improve it— is the explicit task of the engineer. In that sense, the engineers of the past have laid out the foundations of our present quality-of-life. It is the challenging but rewarding task of the engineers today to improve upon this yet.

These are nice and comforting philosophical thoughts which should always be kept in mind. In reality, engineers themselves suffer the almighty control of economics. Engineers work for companies, companies have shareholders and shareholders require profit to improve their individual quality-of-life, not that of the community. Nevertheless, companies can only make profit from products for which there is market. The presence of a market implies that a significant part of the global community benefits from this product. So regardless of economics, engineers aim to serve the community, its facilities, its progress.

1.1 The Growth of the Wireless Communication Market

One of the most prominent areas of recent progress lies in the world of telecommunication. Especially the domain of wireless communication has been marked with a significant growth, the last 15 years. Even though the term wireless communication is very contemporary, it is not at all a new concept. Simply talking to the person next to you, even just looking, is an unmistakeable form of wireless communication. The first form uses sound waves and has a rather low bandwidth for nowadays' applications. The second is in fact a form of wireless optical communication.

The exponential growth of the wireless communication market is a general phenomenon. One of these booming applications is known as the Global Positioning System or GPS. GPS is intended to allow the user to calculate his position relative to earth. The system consists of a constellation of satellites orbiting the earth. The satellites are approximately 20,000 kilometers above the earth and complete 2 orbits per day. A total of 24 GPS satellites make up the constellation and they are positioned to provide 24 hour GPS coverage anywhere on earth.

The worldwide GPS revenues are shown in Fig. 1.1(a). The initial exponential growth is clearly distinguished. With less than $ 1 Billion in 1993 to over $ 16 Billion per year in 2003, the

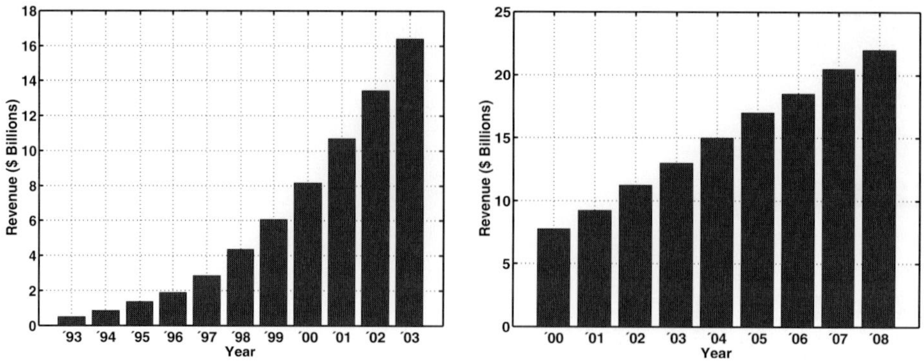

(a) Source[1]. (b) Source[2].

Figure 1.1: *Worldwide GPS equipment revenues. Sources:* [1]*The International Trade Administration, Office of Telecommunications, U.S. Department of Commerce and* [2]*Allied Business Intelligence, Inc...*

average growth is more than 25% per year. The future revenues of worldwide GPS equipment sales are expected to feature a linear increase already up to 2008. A moderate forecast is shown in Fig. 1.1(b).

These studies of the market evolution again highlight the importance of economics in engineering research. One of the main driving forces behind the increasing market and market share, is the reduction of the equipment cost. Especially for luxury goods the price of the product is primordial in our contemplation whether or not to buy it. In the end, the overall sum of these individual decisions determine the size of the market. For GPS specifically, this means that the GPS handsets are offered at a sufficiently low price. As a direct consequence, the manufacturers need to be able to implement these products at a low cost. Hence the receiver front-ends should be available at a low cost as well. This is one of the main motives to investigate the option of integrating the receiver front-end in CMOS.

Fig. 1.2 illustrates the evolution of the GPS 'engine' to the GPS receiver wrist watch, available today. Fig. 1.2(a) plots the area of a typical GPS receiver PCB through the years. In 1993, the area was typically in the order of 200 to 300 cm^2. Putting this bulky GPS receiver on your wrist is comparable to using a wall clock as a wrist watch; quite unpractical. Moreover in 1993 the power consumption of a typical GPS unit was in the order of 2 W as seen in Fig. 1.2(b). This would require the battery of a modern laptop to give the user one or two days of operation. This explains why the price of a single GPS unit was high and the market was very limited. Luckily, progress in technology has reduced the PCB area with about 50% each year, roughly halving the area every other year. The area of the PCB today is only a few cm^2 making it small enough to fit in a watch. The reduction in power consumption follows more or less the same pace: from 2 W in 1993 to less than 50 mW today. To illustrate the consequences, the CASIO wrist watch today, allows GPS radio reception for one day or more depending on the position updating frequency. As area and power consumption gradually scaled down, the corresponding prices have dropped

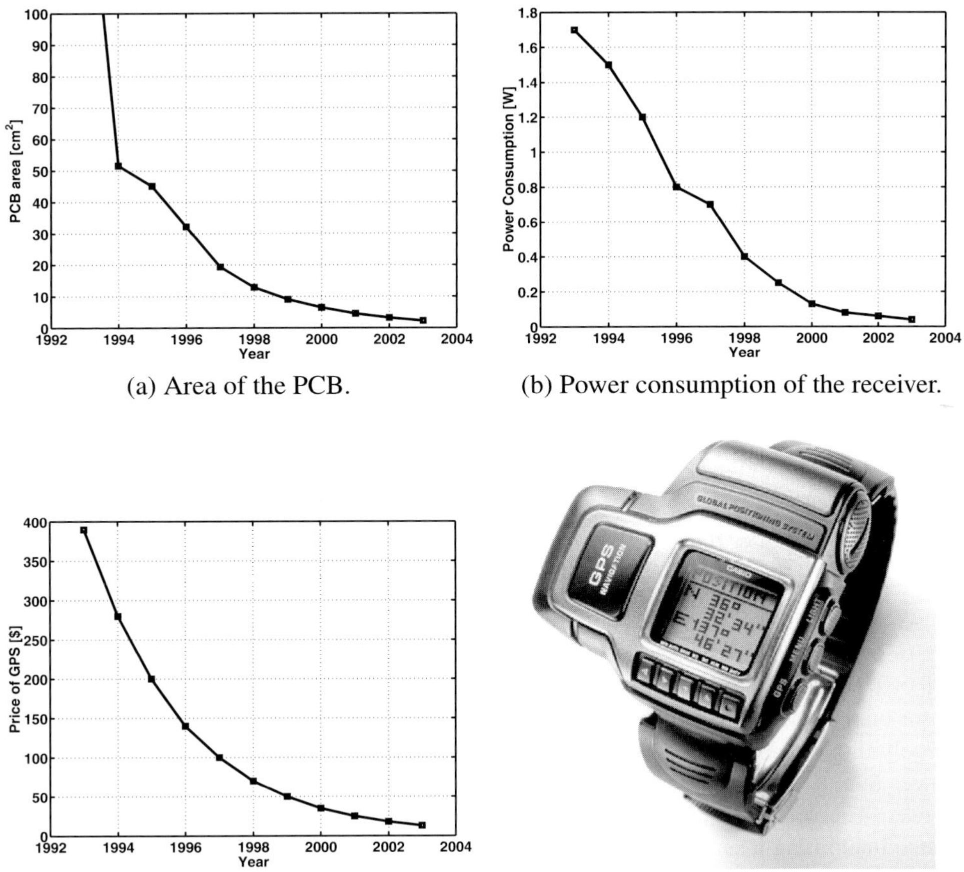

(a) Area of the PCB.

(b) Power consumption of the receiver.

(c) Price of a GPS unit.

(d) GPS receiver in a watch.

Figure 1.2: *Evolution of GPS technology to the GPS in a wrist watch available today.*

as depicted in Fig. 1.2(c). Again, the same exponential descent is distinguished. Also the prices have halved roughly every other year. Today the cost of a GPS receiver is in the order of a few tens of dollars whereas they costed on average close to $ 400 in 1993. This complete evolution is possible due to the continuously increasing level of integration. And integration is the core competence of CMOS.

1.2 Evolution to CMOS RF

CMOS has come a long way since its original invention in the early sixties. First industrial products were introduced in the mid-seventies, more than ten years later. Today, CMOS is the de-facto standard technology for the huge, ubiquitous market of digital IC's. This is why CMOS

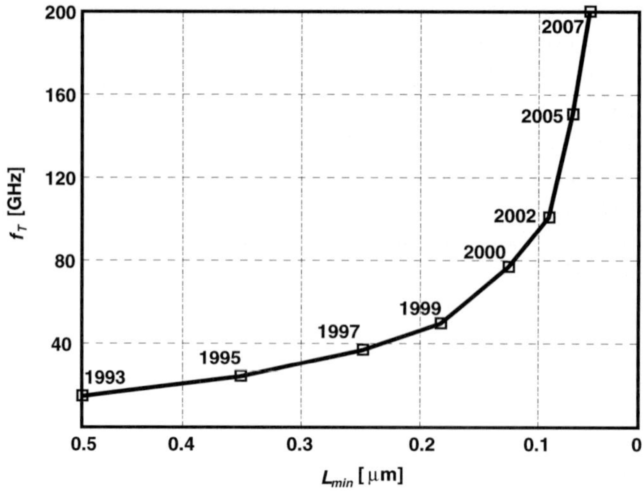

Figure 1.3: *Evolution of the unity-gain frequency, f_T, with decreasing CMOS feature size.*

IC's, also for analog applications, are available at a relatively low price compared to the rare Si bipolar, or even more rare GaAs, SiGe or SOI technologies. Another advantage of using CMOS is that it offers the ability to integrate the RF and analog baseband circuits together with the digital processor on the same die. This has several advantages. The area of the system is reduced since it is largely implemented in a single chip. The power consumption can be reduced by avoiding the power hungry buffers often needed to go off-chip. Unfortunately there are a few severe drawbacks. The very nature of digital processing causes severe switching noise injection in the substrate. Integrating a sensitive RF amplifier with the digital processor on the same die causes the substrate to be shared between both. The presence of this switching noise may severely upset the sensitivity of the front-end. This is one of the reasons, these fully integrated systems are still rare. Even though several precautions can be undertaken to reduce this noise coupling, they are insufficient for high-performance applications. Commercial single-chip CMOS transceivers have nevertheless recently become available for some low-end applications like BlueTooth.

Another drawback results from the removal of external components in the RF front-end. This will often force the migration to different circuit topologies. For instance, in classical super-heterodyne receivers the downconversion was done in several steps. This required intermediate high-Q filtering for image cancellation after all stages. This topology is not interesting for a full CMOS wireless receiver since these high quality mixers have to be implemented externally and they would excessively increase the overall system cost. Therefore, two other topologies have been devised which don't need these high-Q filters: the zero-IF receiver [Abi95, Raz97] and the low-IF receiver [Cro98]. Both are direct conversion receivers since they don't require a second downconversion step.

In spite of these persistent obstacles for providing a complete integrated transceiver, CMOS is still becoming an attractive candidate for RF applications, owing to its inherent low cost.

Moreover, as technologies scale down to increase the speed and reduce the power consumption of digital IC's, the f_T (unity current gain frequency) increases as well, which intrinsically allows higher frequency operation. The f_T is plotted versus the feature size of the CMOS technology in Fig. 1.3. The years in which the technology has become or will become available is indicated as well. It is commonly believed that CMOS RF receivers are feasible for frequencies roughly below $f_T/10$. A nice example is the 2 V CMOS cellular transceiver front-end in [Ste00a]. It has been implemented in an early 0.25 μm technology corresponding to a maximum f_T of about 30 GHz. The transceiver is fully integrated and operates at 1.9 GHz, more or less confirming the above rule-of-thumb. The nowadays emerging standard 90 nm CMOS technologies offer an f_T close to 100 GHz. This implies that these technologies will allow fully integrated RF receivers up to as much as 10 GHz. This is more than sufficient for the 5 - 6 GHz Wireless LAN standards IEEE 802.11a and HIPERLAN. Naturally this rule-of-thumb should be used with care since a lot depends on the system specifications. For the transmitter part the rule-of-thumb is more or less equivalent except for the power amplifier. There, the lower breakdown voltage of the smaller technologies counteracts the increased f_T and f_{max}. Depending on the required output power it may be necessary to implement the power amplifier externally in a more dedicated technology.

1.3 CMOS, RF and ESD

For any economically viable product, reliability is a serious issue. This applies to the IC world as well. In this context, reliability refers to several distinct topics. A product is reliable when it does what it's supposed to do under all *normal* circumstances. For an IC this means that the circuit needs to operate within specifications under all possible conditions, with respect to temperature, humidity, etc.. Naturally there is an inherent offset between two IC's which is due to inevitable variations in process parameters. The yield refers to the percentage of chips working within the specification boundaries. A large yield is indispensable for a reliable product.

Another field of reliability requires the circuit to remain functional under normal and abnormal human or machine handling. This leads us immediately to the circuit's affinity for Electro-Static Discharge or ESD. Any IC that hits the market is assumed to have some built-in immunity to ESD. The amount of protection required depends on the application. An IC pin connecting directly to the outside world is much more susceptible for ESD events than a pin only used in the product's interior. For an RF receiver, the input can be connected immediately to an external antenna. This directly exposes the IC to a human discharging through the conductive antenna. Though this is an extreme example, each chip is exposed to contact with machines and or humans during moving, packaging and assembling. During any of these actions they can be exposed to ESD. This is why companies assembling IC's into an intermediate or final product usually require them to have ESD protection.

In fact, one of the final bottlenecks for introducing CMOS RF circuits to the market is their susceptibility to ESD. It is due to both gate oxide breakdown and junction degradation related problems. They are further aggravated by the decreased oxide thickness and increased doping levels in the scaled down technologies. Most CMOS ESD-protection structures (e.g. as they are used in digital CMOS) have parasitics that can be detrimental for the RF performance.

One of the blocks that suffer the most from the ESD requirements is the low-noise amplifier. It is the first block in any integrated receiver. The LNA aims to amplify the RF input signal as much as possible without adding a significant amount of noise. The specified minimum signal level of the application should be lifted above the noise-floor of the subsequent mixer stage. The low-noise performance of the LNA makes it extremely susceptible to any input parasitic. Resistors of course but also capacitive and inductive parasitics may seriously degrade the noise performance. The ESD-protection circuits have both resistive and capacitive parasitics which inevitably degrade the noise figure of the LNA. The impact of the ESD-constraints on the design of RF low-noise amplifiers will be investigated thoroughly in this book. The sensitivity requirement of modern GPS receivers makes the GPS LNA an ideal demonstrator for the design of ESD-protected LNA's. It will be investigated how classical ESD protection circuits and devices affect the RF performance. New topologies will be proposed that overcome the performance limitations imposed by the classical ESD circuits. These topologies require a rigorous co-design of both LNA and ESD protection.

1.4 Outline of this Book

- Chapter 2 gives a general overview of the area of low-noise amplification in CMOS wireless receivers. It starts with a brief introduction of the most important RF concepts. The device models that will be used in hand calculations are explained. These hand calculations will be applied throughout the text to explain the behavior of the circuits. The models are sufficiently simple to keep the resulting equations manageable. They are sufficiently complex to accurately describe the important phenomena. The parameters of the models can be adapted to simulation results in order to improve local or global accuracy. These models and calculations will be used repetitively to generate contour plots of different design and performance parameters. The plots aim to give intuitive insight in the behavior of the circuit. An extended MOS model is introduced that has been used in conjunction with the numerical simulators. After a general introduction on noise, the LNA function is described within the receiver chain. The main design criteria and performance requirements are derived. In conclusion of this chapter, the most common LNA topologies in CMOS are classified and discussed. Some specific and interesting designs, published in open literature but falling beyond the above classification are clarified as well.

- Chapter 3 gives a quick survey of the different ESD tests and standards and the various ESD protection devices and topologies. One of the most commonly used standards to give an indication of the protection level is based on the Human Body Model test (HBM). The amount of protection is indicated by the HBM voltage the circuit is able to withstand. The standard level of protection for an IC is 2 kV.

- Chapter 4 will discuss in detail one of the most interesting topologies for RF LNA's, the common-source LNA with inductive source degeneration. It is so interesting because it enables an extremely low NF and high gain which is mandatory for many of today's wireless receivers requiring a good sensitivity (GPS, GSM, DCS1800, UMTS, etc.). The influence of all relevant parasitic components will be discussed. Their impact on the different performance criteria (noise, gain, linearity, matching and stability) will be explained by means of the relevant design equations. Numerical evaluations are based on a demonstrator design of a 1.5 GHz LNA in a 0.25 μm CMOS technology.

- Chapter 5 is dedicated to the introduction of new RF-ESD co-design methodologies which are able to satisfy both RF and ESD requirements for high-frequency LNA's. The discussion is based on the CS LNA with source degeneration inductor but is extended at the end towards other topologies. The chapter starts by reviewing the frequency limitations of the classical CS LNA. Two different RF-ESD co-design strategies are introduced which overcome these limitations. The first technique is based on the use of a Π-type input matching network. The second solution uses an on-chip inductor to drain the ESD charges. The different methods are evaluated numerically with a demonstrator design at 5 GHz in the same 0.25 μm technology.

- In Chapter 6, the design, layout and measurements of three LNA prototypes are discussed. All circuits have been provided with ESD-protection. The first chip is a low-noise amplifier for the L2 GPS band at 1.23 GHz. It has been implemented in a 0.25 μm technology. A second low-noise amplifier has been designed and integrated within a complete L1 GPS receiver front-end. It has been implemented in the same technology. The last design targets 5 GHz wireless LAN applications. The circuit features an integrated ESD-protecting inductor.

Chapter 2

Low-Noise Amplifiers in CMOS Wireless Receivers

2.1 Introduction

This chapter aims to welcome the reader to the world of low-noise amplification in wireless receivers. The most important RF concepts are introduced in Section 2.2. These concepts include the quality factor of reactive elements, different types of matching, power gain and distortion. Topics that will return and gain significance in various discussions further on.

Section 2.3 introduces the device models that will be used in hand calculations. Hand calculations will be applied throughout the text to give intuitive insight in the behavior of different circuits and circuit aspects. The models have been fit in advance to numerical simulation results. Also an extended MOS model is introduced that has been used in conjunction with the numerical simulators. The most common noise sources in CMOS IC's are discussed together with their physical origin in Section 2.4. In Section 2.5, the LNA is described in its function and functionality within the receiver chain. The coherence and mutual dependence of the LNA with the other receiver blocks is investigated. Based on that, the main design criteria and performance requirements are derived.

To conclude this chapter, the most common LNA topologies in CMOS are classified and introduced with a simple —but not irrelevant— performance model in Section 2.6. Already a swift comparison can be made. Some specific and interesting designs, published in open literature but falling beyond the above classification are clarified briefly.

2.2 Some Important RF Concepts

2.2.1 Quality Factor of Reactive Elements and Series-Parallel Transformation

A few concepts that will reoccur often are the resonance, quality factor and series-parallel transformation of reactive elements. For a purely reactive element the current through the element is

90 degrees out of phase with the voltage over it. Hence no power is consumed in the element. Naturally we are talking about inductors and capacitors where the currents are respectively lagging and leading the voltage by 90 degrees. In real life however a purely reactive element does not exist and some power dissipation is always present. Moreover if there is power dissipation, there is a resistor and resistors give rise to thermal noise whereas reactive elements are completely noiseless. Consequently a means is needed to describe the '*purity*' of a reactive element.

This means is known as the quality factor Q of the reactive element. It is defined by:

$$Q \triangleq \frac{\text{average reactive power}}{\text{average power dissipated}}. \tag{2.1}$$

For a simple inductor or capacitor with a series resistor R_s this expression becomes

$$Q_L = \frac{\omega L}{R_s} \qquad \text{and} \qquad Q_C = \frac{1}{\omega C R_s} \tag{2.2}$$

respectively. This can be rewritten in one formula:

$$Q = \frac{X_s}{R_s}, \tag{2.3}$$

where X_s is the reactance of either inductor or capacitor at the given frequency. For an inductor or capacitor with a parallel resistor R_p, the quality factor is found as

$$Q = \frac{R_p}{B_p}, \tag{2.4}$$

where B_p is the susceptance of the inductor or capacitor at the given frequency.

A quality factor can also be constructed for a resonant RLC network. Consider a series RLC tank. The tank is characterized by its resonance frequency

$$\omega_r = \frac{1}{\sqrt{LC}}, \tag{2.5}$$

and by its quality factor

$$Q = \frac{\omega_r L}{R} = \frac{1}{\omega_r C R}. \tag{2.6}$$

This means that the Q defined by (2.2) is equivalent to the Q of a series RLC tank with resonant frequency $\omega = \omega_r$. Due to the series-parallel duality this equivalence applies also for a parallel RLC tank and equation (2.4).

For a simple RLC tank, the Q-factor has yet another meaning. Consider the impedance of a parallel tank. The quality factor of a RLC tank is related to the sharpness of the impedance peak, or mathematically:

$$Q = \frac{2\pi \text{BW}}{\omega_r}, \tag{2.7}$$

where BW is the total (left and right) -3 dB bandwidth of the impedance magnitude centered around ω_r. Again, by duality (2.7) is also valid for a series tank but one needs to take the admittance bandwidth.

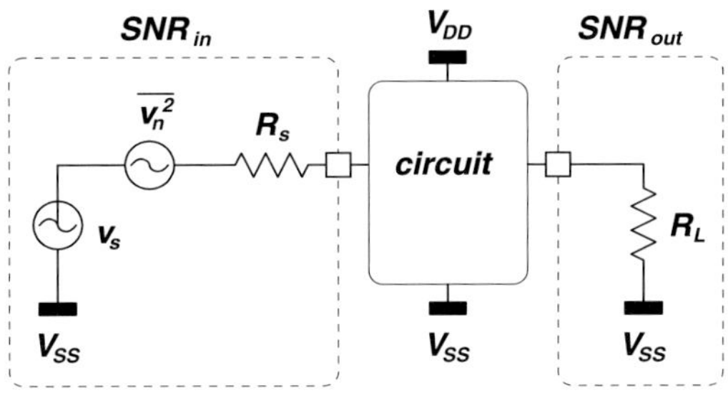

Figure 2.1: *Input and output Signal-to-Noise Ratio.*

2.2.2 SNR and Noise Figure

The SNR or Signal to Noise Ratio gives a measure for the purity of a signal. The definition is quite simple:

$$\mathrm{SNR} = \frac{\text{Available Signal Power}}{\text{Available Noise Power in Signal Bandwidth}}. \tag{2.8}$$

For instance the SNR_{in} of the signal source represented in Fig. 2.1 is

$$\mathrm{SNR}_{in} = \frac{P_{av,s}}{P_{av,n}} = \frac{\frac{v_s^2}{4R_s}}{kT\Delta f}, \tag{2.9}$$

where Δf is the signal bandwidth. Rewriting equation (2.9) as

$$\mathrm{SNR}_{in} = \frac{v_s^2}{4kTR_s\Delta f} \tag{2.10}$$

shows that it doesn't matter whether the ratio of squared voltages power or squared current is taken since both noise and signal have the same conversion factor, determined by the respective node impedance.

An ideal amplifying block operating on the signal will amplify both signal and noise equally and will not alter the SNR. However, any real-life —non-ideal— block will decrease the SNR since the block will add some noise to the signal. Mathematically this is expressed by the noise factor of the block:

$$F = \frac{\mathrm{SNR}_{in}}{\mathrm{SNR}_{out}} = \frac{\frac{P_{av,s}}{P_{av,n}}}{\frac{G \cdot P_{av,s}}{G \cdot P_{av,n} + G \cdot P_{n,eq}}}, \tag{2.11}$$

where $G \cdot P_{n,eq}$ is the excess noise power at the output and $P_{n,eq}$ is this power referred to the input. This can be simplified to

$$F = \frac{P_{av,n} + P_{n,eq}}{P_{av,n}}. \tag{2.12}$$

This means the noise factor is the total equivalent input noise power divided by the noise power of the source. Or equivalently, the noise factor is the total output noise divided by the output noise resulting solely from the noise power of the input source. The noise figure is used much more often than the noise factor. It is related to the noise factor according to

$$\mathrm{NF} = 10\log(F). \tag{2.13}$$

Since F can be any number between 1 and ∞, NF is bounded by 0 and ∞. Noise figures lower than 0 should arouse serious suspicion since any sort of selective noise absorber has yet to be invented!

Now consider the specific case of the low noise amplifier. The LNA is usually driven by a 50 Ω source which can be either the impedance of the receive antenna or the output impedance of a band selecting SAW-filter. Consider the first case. The input SNR is given by

$$\mathrm{SNR}_{in} = \frac{v_s^2}{4kT_{\mathrm{eff}}R_s\Delta f} \tag{2.14}$$

(2.10) where $R_s = 50\,\Omega$ en T_{eff} is the effective noise temperature of the antenna. For the common case where the radiation resistance far exceeds the resistive losses in the antenna leads, T_{eff} is the average noise temperature seen by the antenna. It can be described by

$$T_{\mathrm{eff}} = \int_0^{4\pi} T(\Psi)G_A(\Psi)d\Psi, \tag{2.15}$$

where Ψ is the solid angle expressed in steradians, $T(\Psi)$ is the temperature at solid angle Ψ and $G_A(\Psi)$ is the antenna gain for solid angle Ψ. This temperature is largely dependent on the *view* of the antenna. For a GPS receiver for instance, T_{eff} will be very low at night looking into the sky with a temperature of only a few tens of Kelvin depending on the quality of the antenna. However in daylight, looking at the sun, the effective noise temperature will be much higher. In order to have a fixed noise factor for the LNA independent of the noise temperature of the antenna, F_{LNA} is defined with a fixed source noise temperature equal to the physical room temperature:

$$F_{LNA} = \frac{kT_r\Delta f + P_{n,eq}}{kT_r\Delta f} = 1 + \frac{P_{n,eq}}{kT_r\Delta f}. \tag{2.16}$$

Since most noise sources are proportional to the physical temperature, equation (2.16) shows that F_{LNA} should be independent of the actual room temperature during the measurements.

2.2.3 Impedance Matching, Power Matching, Noise Matching

Impedance matching is a term which is used frequently in the area of transmission lines. A transmission line is characterized by a characteristic impedance Z_c. Suppose the line is terminated with an impedance Z. A voltage wave V^+ travelling along the line will be partially reflected at the end of the line depending on the termination impedance. The reflected voltage V^- is given by

$$V^- = \Gamma V^+ \tag{2.17}$$

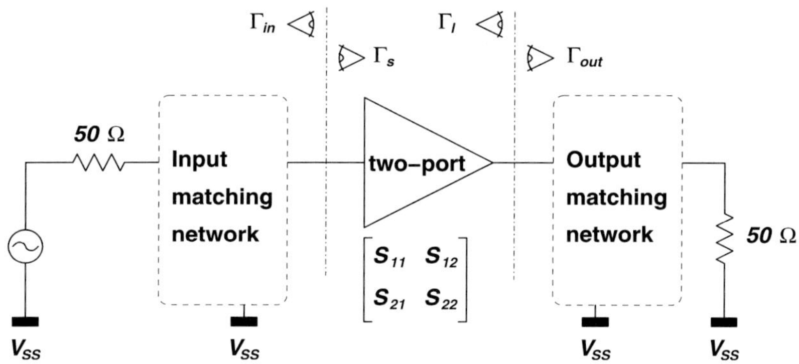

Figure 2.2: *An arbitrary two-port with lossless input and output matching networks.*

where

$$\Gamma = \frac{Z - Z_c}{Z + Z_c}. \tag{2.18}$$

Note that Γ is a complex number comprising both the amplitude ratio and the phase turn. If $Z = Z_c$ then $\Gamma = 0$ and no reflection occurs.

Power matching is in essence not related to impedance matching. The origin of power matching lies in the fundamental quest for energy efficiency. Suppose a voltage source (voltage V_S) with a source impedance Z_S drives a load impedance Z_L. The question is what value of Z_L maximizes the power dissipation in the load. It can easily be shown that this is achieved when

$$Z_L = Z_S^*, \tag{2.19}$$

with a maximum dissipated power in the load calculated as

$$P_{max} = \frac{V_S}{4\Re(Z_S)} \triangleq P_{av}. \tag{2.20}$$

This is also called the available source power.

Noise matching is completely unrelated to both previous types of matching. The origin here is the quest for good SNR and hence low noise figure. For a given two-port a noise match is obtained when the impedance of the source driving the two-port minimizes the noise figure of the resulting system. Referring to Appendix A this is achieved when $Z_S = Z_{opt}$.

In what follows the word '*matching*' must always be interpreted as '*impedance matching*' unless specifically stated otherwise.

2.2.4 Transducer Power Gain, Operating Power Gain and Available Power Gain

The concept of power gain of a two-port is not unambiguous. Several kinds of power gain are defined. Consider an arbitrary two-port as depicted in Fig. 2.2.

Transducer Power Gain or G_T is defined as follows:

$$G_T = \frac{\text{Power absorbed by the load}}{\text{Available power of the source}}. \tag{2.21}$$

Referring to Fig. 2.2, this can be rewritten as

$$\begin{aligned} G_T &= \frac{1-|\Gamma_s|^2}{|1-\Gamma_{in}\Gamma_s|^2}\cdot|S_{21}|^2\cdot\frac{1-|\Gamma_l|^2}{|1-S_{22}\Gamma_l|^2} && (2.22)\\ &= \frac{1-|\Gamma_s|^2}{|1-S_{11}\Gamma_s|^2}\cdot|S_{21}|^2\cdot\frac{1-|\Gamma_l|^2}{|1-\Gamma_{out}\Gamma_l|^2}, && (2.23)\end{aligned}$$

where

$$\Gamma_{in} = S_{11} + \frac{S_{12}S_{21}\Gamma_l}{1-S_{22}\Gamma_l} \tag{2.24}$$

represents the reflection coefficient of the one-port constructed by the amplifier connected to the load Γ_l, and

$$\Gamma_{out} = S_{22} + \frac{S_{12}S_{21}\Gamma_s}{1-S_{11}\Gamma_s}. \tag{2.25}$$

is the equivalent representation of the output reflection coefficient.

The transducer power gain is most frequently used since the available source power is a given and the power in the load is what should be maximized.

Operating Power Gain or G_p is probably the most obvious definition. It is given by:

$$G_p = \frac{\text{Power absorbed by the load}}{\text{Power absorbed at the input}}. \tag{2.26}$$

Rewriting this in function of the reflection coefficients, yields

$$G_p = \frac{1}{1-|\Gamma_{in}|^2}\cdot|S_{21}|^2\cdot\frac{1-|\Gamma_l|^2}{|1-S_{22}\Gamma_l|^2}. \tag{2.27}$$

Since this definition represents the output power normalized to the absorbed input power, it is independent of the actual equivalent source impedance represented by Γ_s.

Available Power Gain or G_{av} is defined as

$$G_{av} = \frac{\text{Available output power}}{\text{Available power of the source}}. \tag{2.28}$$

As a function of matching coefficients, this becomes

$$G_{av} = \frac{1-|\Gamma_s|^2}{|1-S_{11}\Gamma_s|^2}\cdot|S_{21}|^2\cdot\frac{1}{1-|\Gamma_{out}|^2}. \tag{2.29}$$

Since the available power gain refers to the available output power it is independent of the actual equivalent load impedance represented by Γ_l.

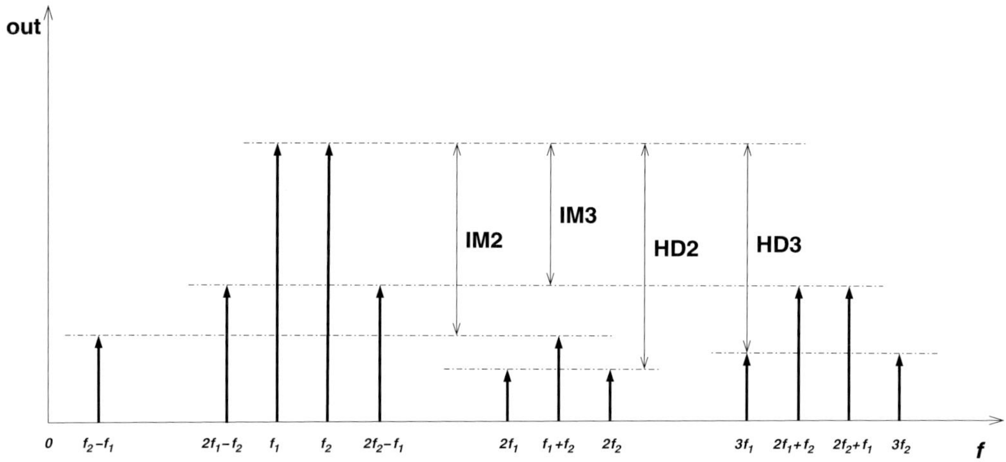

Figure 2.3: *Output tones in a two-tone test for a system with second and third order distortion.*

The available output power is always larger than the absorbed power by the load, therefore $G_T \leq G_{av}$. Similarly, the power absorbed at the input is always smaller than the available source power, hence $G_T \leq G_p$.

In the remainder of this work we will only use the transducer power gain which is shortened to power gain. In most cases the power gain of the LNA is simply equal to S_{21} unless specifically mentioned otherwise. The amplifiers are designed such that input and output impedance are sufficiently close to $50\ \Omega$ to justify this simplification. In other words Γ_s, Γ_{in}, Γ_l and Γ_{out} are sufficiently close to zero.

2.2.5 Intermodulation Distortion

Basically two kinds of non-linearities can be distinguished: weak non-linearities and hard non-linearities. The first kind can be described by a Taylor series and can be approximated with arbitrary accuracy by simply taking sufficient terms in the expansion. An example of a weak non-linearity is the $i_{ds} - v_{gs}$ relation for a MOS transistor in saturation. Hard non-linearities, for instance clipping, can not be described with a finite Taylor expansion. Typical for hard non-linearities is that almost no non-linearity is present for very small input amplitudes but all of a sudden the system behaves extremely non-linear (for instance when clipping starts). In the further analysis all non-linearities are considered to be weak non-linearities.

The linearity of circuits is usually investigated by means of harmonic distortion analysis or intermodulation distortion analysis. The first one assumes a sine wave is applied to the input. The fundamental and harmonics at the output are studied. For intermodulation, two tones are applied at the input and the intermodulation terms together with the fundamental tones are investigated.

Fig. 2.3 shows the output tones for a system with second and third order distortion. Suppose two tones are applied at frequencies f_1 and f_2. Besides the fundamental tones, the output also shows second and third order harmonics and second and third order intermodulation products. The second order intermodulation gives rise to tones at $\pm(f_1 - f_2)$ and at $\pm(f_1 + f_2)$ as illustrated in Fig. 2.3. If f_1 and f_2 are located around the carrier f_c, then $f_1 - f_2 \approx 0$ and $f_1 + f_2 \approx 2f_c$. The first will be rejected by the DC-offset compensation and the second will usually be filtered out. Moreover, second order intermodulation terms are often very low due to differential implementations such that second order terms appear as common mode. Still, sufficient care needs to be taken in the receiver design since out of band signals may have a second order intermodulation term falling in the band of interest. The CMRR should be good enough to avoid signal degradation as a result of these signals.

Third order intermodulation will cause tones at frequencies $\pm(2f_1 - f_2)$, $\pm(2f_2 - f_1)$ and $\pm(2f_1 + f_2)$. When the applied tones are close to the carrier, the last intermodulation tone will be close to $3f_c$ and be filtered out but the first two will be within the band of interest. Since they are not linearly correlated with the input signal they can be considered as noise disturbing the signal. This is why the definition of SNR discussed in Section 2.2.2 is extended to SNDR, the signal to noise and distortion ratio:

$$\text{SNDR} = \frac{P_{av,s}}{P_{av,n} + P_{im}}, \tag{2.30}$$

where P_{im} is the combined power of the in band intermodulation signals.

The ratio of the amplitude of the third order intermodulation signals and the amplitude of the fundamental signal is called IM3. Consequently IM3 increases with the square of the input signal amplitude. Consider a system described by

$$y(t) = f(x(t)). \tag{2.31}$$

Performing a Taylor expansion of $y(t)$ yields

$$y(t) = a_0 + a_1 x(t) + a_2 x^2(t) + a_3 x^3(t) + O(x^4(t). \tag{2.32}$$

where

$$a_k = \frac{1}{k!}\frac{d^k y}{dx^k}. \tag{2.33}$$

Two tones are applied at the input:

$$x(t) = U\sin(\omega_1 t) + U\sin(\omega_2 t). \tag{2.34}$$

IM3 is found as

$$\text{IM3} = \frac{3}{4}\left|\frac{a_3}{a_1}\right| \cdot U^2, \tag{2.35}$$

where U is the input signal amplitude. Note that (2.35) can be used for any weakly non-linear circuit. The input amplitude for which IM3 $= 1$ is called the input referred third order intermodulation intercept point or IIP3:

$$\text{IIP3} = \sqrt{\frac{4}{3}\left|\frac{a_1}{a_3}\right|}. \tag{2.36}$$

If a_1 is very large compared to a_3 and neglecting a_2 then IIP3 will be large and the output will be a linearly scaled version of the input signal for a wide input range.

2.3 The Deep Sub-Micron MOS Transistor at Radio Frequencies

2.3.1 MOS Model for Hand Calculations

The models and equations discussed in this section will be used throughout this work for hand calculations. The model is quite similar to [HSp01] MOS level 3 . The drain-source current of a NMOS in saturation is described by

$$I_{DS} = \frac{\mu C_{ox}}{2} \frac{W}{(L - \Delta L)} \frac{V_{GST}^2}{(1 + \Theta V_{GST})}, \tag{2.37}$$

where μ is the mobility, $C_{ox} = \epsilon_{ox}/t_{ox}$, Θ models the mobility degradation due to both longitudinal electric field (velocity saturation) and transverse electric field and V_{GST} symbolizes $V_{GS} - V_T$ in order to reduce the complexity of the expressions. The factor $L - \Delta L$ takes into account the channel length modulation. $\Delta L/L$ is — within a limited range — proportional to V_{DS} and therefore, (2.37) can be rewritten as

$$I_{DS} = K\frac{W}{L} \frac{V_{GST}^2}{(1 + \Theta V_{GST})} \frac{1}{(1 - \Lambda V_{DS})}, \tag{2.38}$$

where K, Θ and Λ are extracted from simulations (HSpice or Eldo) by means of the MOSCAL tool [Van02a]. Equation (2.38) describes the behavior of the transistor well within a selected region of operation. Naturally as this region is increased the model becomes increasingly inaccurate. Hence for very fine calculations — for instance during design optimization —, the design space needs to be split up into several smaller regions with its own set of describing parameters.

The small signal parameters used in the hand calculations can be derived from (2.38). The transconductance is found by differentiating I_{DS}:

$$g_m \triangleq \frac{\partial I_{DS}}{\partial V_{GS}} = 2K\frac{W}{L} V_{GST} \frac{1}{(1 + \Theta V_{GST})} \left(\frac{1 + \frac{\Theta}{2} V_{GST}}{1 + \Theta V_{GST}} \right) \frac{1}{(1 - \Lambda V_{DS})} \tag{2.39}$$

The cut-off pulsation neglecting C_{gd} is now found as

$$\omega_T = \frac{g_m}{C_{gs}} = \frac{\mu V_{GST}}{L^2} \frac{1}{(1 + \Theta V_{GST})} \left(\frac{1 + \frac{\Theta}{2} V_{GST}}{1 + \Theta V_{GST}} \right) \frac{1}{(1 - \Lambda V_{DS})} \tag{2.40}$$

The finite output resistance due to the channel length modulation is approximated by

$$r_{ds} = \frac{1}{\Lambda I_{DS}}. \tag{2.41}$$

This can easily be understood from (2.38) if the factor $1/(1-\Lambda V_{DS})$ is replaced with $(1 + \Lambda V_{DS})$ which is justified if $\Lambda V_{DS} \ll 1$. Note that even though Λ is inversely proportional to the effective channel length, this dependence can be ignored since all transistors feature the minimal length.

NMOS

K_n	V_{Tn}	Θ_n	Λ_n	α	α_{gd}	α_{gb}	$\alpha_{db} = \alpha_{sb}$ $V_{DB} = 0.5$ V	$\alpha_{db} = \alpha_{sb}$ $V_{DB} = 1.5$ V	γ	δ
[μA/V^2]	[V]	[V^{-1}]	[V^{-1}]	[ ]	[ ]	[ ]	[ ]	[ ]	[ ]	[ ]
192	0.52	4.15	0.07	0.83	0.23	0.16	0.68	0.53	2	4

PMOS

K_p	$\|V_{Tp}\|$	Θ_p	Λ_p	α	α_{gd}	α_{gb}	$\alpha_{db} = \alpha_{sb}$ $V_{DB} = 0.5$ V	$\alpha_{db} = \alpha_{sb}$ $V_{DB} = 1.5$ V	γ	δ
[μm]	[μA/V^2]	[V]	[V^{-1}]	[ ]	[ ]	[ ]	[ ]	[ ]	[ ]	[ ]
55	0.50	3.87	0.07	0.83	0.23	0.16	0.68	0.53	1	2

COMMON

L_{eff} [μm]	t_{ox} [nm]
0.2	5.5

Table 2.1: *Hand calculation parameters for the NMOS and PMOS in the 0.25 μm CMOS technology of Kawasaki Microelectronics (extracted for $V_{GS} - V_T$ values between 0.1 and 0.3 V).*

Unless specifically stated otherwise, all illustrated calculations have been done based on the hand calculation parameters in Table 2.1. Hereby, α is defined as

$$\begin{aligned} \alpha &= \frac{g_m}{g_{d0}} \\ &\approx \frac{1}{n} \triangleq \frac{g_m}{g_m + g_{mb}}, \end{aligned} \tag{2.42}$$

where g_{d0} is the drain-source conductance at zero V_{DS}. Parameters α_{xy} are defined as

$$\alpha_{xy} = \frac{C_{xy}}{C_{gs}}, \tag{2.43}$$

and γ and δ represent the excess noise factors discussed in Section 2.4.2.

2.3.2 Linearity of the short-channel MOS transistor

Since for an LNA, the main non-linearity problem is the 3rd order intermodulation, this subsection will evaluate the intermodulation performance of a MOS transistor by means of the IV3. This is the gate-source voltage amplitude for which the intermodulation drain current intercepts the fundamental drain current. In principle it is identical to the IIP3 —as it was introduced in Section 2.2.5— in as far as the gate-source voltage amplitude is the actual input and no conversion for input reference is required. The symbol IV3 is used here for the more general case where the input signal is different which will be the case in the amplifiers discussed further on. This

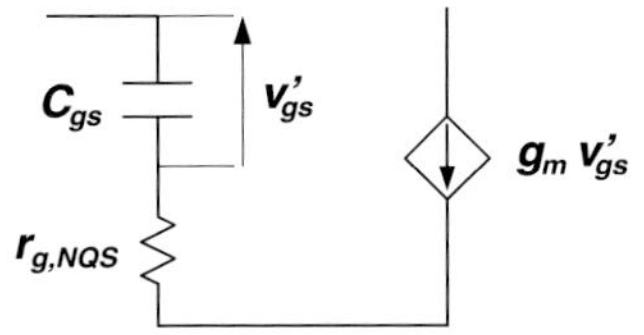

Figure 2.4: *Non-Quasi Static model for the delay in the channel charge buildup.*

will avoid confusion with the actual input referred intercept point or IIP3. The small signal input is $v_{gs}(t)$ and the output of the transistor is the current $i_{ds}(t)$. The total output current I_{DS} of a NMOS is described by (2.38). This reduces to

$$I_{DS} = K\frac{W}{L}\frac{V_{GST}^2}{(1+\Theta V_{GST})}. \tag{2.44}$$

where V_{DS} is assumed constant. This assumption is usually justified as shown mathematically in [Jan01]. Now the small signal current i_{ds} can be written as

$$i_{ds} = K\frac{W}{L}\left(\frac{(V_{GST}+v_{gs})^2}{(1+\Theta(V_{GST}+v_{gs}))} - \frac{V_{GST}^2}{(1+\Theta V_{GST})}\right). \tag{2.45}$$

This function can be expanded in a Taylor series which after some calculation, similar to the general derivation in Section 2.2.5, yields the following expression for IV3:

$$\text{IV3} = \sqrt{\frac{4}{3}\left|\frac{a_1}{a_3}\right|} \tag{2.46}$$

$$= \sqrt{\frac{4}{3}\frac{V_{GST}(2+\Theta V_{GST})(1+\Theta V_{GST})^2}{\Theta}}, \tag{2.47}$$

where IV3 is expressed in Volt amplitude. For any NMOS in saturation, the gate-source intercept voltage is given by (2.47).

2.3.3 Non-Quasi Static Model

The classical quasi static model of the MOS transistor behavior assumes that any change in charge at the gate is instantly reflected with an equal but opposite amount of charge in the channel. However, in reality there will always be a delay in the channel charge buildup. The physics of the MOS transistor tells us that the channel is built by means of inversion. Remember the behavior of the NMOS capacitor where the channel depletion starts when the gate voltage is increased above 0 V. Above V_T, electrons will be drawn from the bulk material creating an excess of inversion carriers in the channel. Considering this, it is intuitively clear that the process of adding an extra electron to the channel has a finite time constant.

(a) Series resonance. (b) Parallel resonance.

Figure 2.5: *The capacitive input of the MOS transistor tuned out with an inductor.*

This effect has been described and modelled by Y. Tsividis in [Tsi87, Jan99a]. A simplified model valid in strong inversion and within the long-channel approximation yields the following time constant associated with C_{gs}:

$$\tau_{gs} = \frac{C_{gs}}{5g_m} = \frac{1}{5\omega_T}. \tag{2.48}$$

Consequently, this delay effect can be modelled by adding a resistor $r_{g,NQS}$ in series with C_{gs}:

$$r_{g,NQS} = \frac{1}{5g_m}. \tag{2.49}$$

This model is illustrated in Fig. 2.4.

The frequency corresponding to this time constant is $5 \times f_T$ so one would think that this effect is not important at realistic operating frequencies much smaller than f_T. However in bandpass applications, the input capacitance may be tuned out with a series inductor. This means that for a given input current the voltage over C_{gs} is cancelled by the equal but opposite voltage over the inductor. The input impedance of the transistor is now purely resistive as shown in Fig. 2.5(a):

$$Z_{in,s} = r_{g,NQS} = \frac{1}{5g_m}. \tag{2.50}$$

Similarly, when the input capacitance is tuned out with a parallel inductor, the input is again purely resistive as demonstrated in Fig. 2.5(b):

$$Z_{in,p} \approx \frac{1}{\omega_r^2 C_{gs} r_{g,NQS}} = \frac{5f_T^2}{g_m f_r^2} \tag{2.51}$$

where f_r is the resonance frequency.

In short-channel MOS transistors the value of the non-quasi static resistor is still under discussion. It is generally assumed that the proportionality with g_m remains but the constant might be changed. Therefore (2.49) is rewritten as

$$r_{g,NQS} = \frac{1}{\kappa g_m}, \tag{2.52}$$

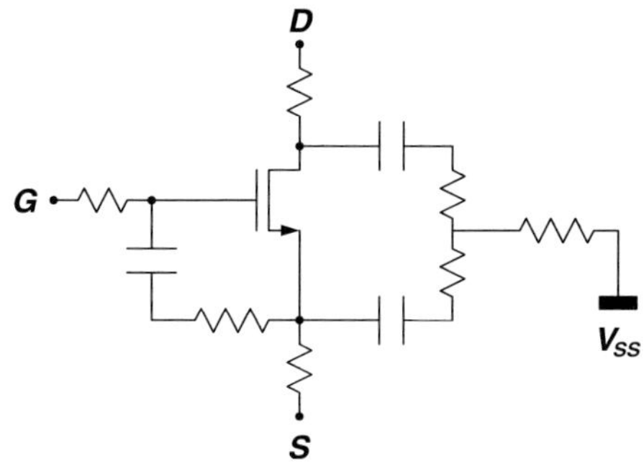

Figure 2.6: *NMOS transistor with 10 additional components for more accurate simulations.*

where κ represents the Elmore constant. The Elmore constant is usually represented by ϵ but κ is used here to avoid confusion with the permittivity. This NQS model is also implemented in MOS model BSIM3V3. All numerical examples in this text will assume $\kappa = 5$ unless specifically stated otherwise [Enz02].

2.3.4 Extended MOS Model for Simulation

Most simulations are performed using HSpice or Eldo in combination with Berkeley MOS model BSIM3V3 (level 49 in [HSp01] and level 53 in [Eld01]). Even though this model is far more complex than the one used for hand calculations (cf. Section 2.3.1), it is still lacking intrinsic accuracy for RF simulations. The designer however has a lot of options to solve this. An NMOS transistor is shown in Fig. 2.6 where 10 extra components have been added to better describe the behavior at high frequencies and more accurately predict the noise behavior.

If the NQS effect is not included in the MOS model that is used it can be modelled by placing the equivalent parallel resistor in parallel with C_{gs} given by

$$R_{p,NQS} \approx \frac{1}{\omega_0^2 C_{gs} r_{g,NQS}} = \frac{5f_T^2}{g_m f_0^2}. \tag{2.53}$$

This model is only valid at frequency f_0 and should be used with care. An extra coupling capacitance is added in series with $R_{p,NQS}$ in order not to disturb the operating point of the circuit.

Fig. 2.6 also shows the series resistors for the gate, source and drain region. The gate resistor represents the resistance of the poly gate. Taking it into account is important for accurate noise simulations. The same goes for the source resistor which models the resistance of the n+ source region. The drain resistor can also be important for instance in switched power amplifiers where it will increase the on-resistance of the switch.

Both at drain and source a capacitor is added to represent their respective junction capacitances. They are resistively coupled to the bulk node. Finally a resistor is added representing the resistance from the bulk node to the actual bulk contact. This resistor is relatively large since it is formed in a high ohmic p-well or n-well region.

Figure 2.7: *Noise voltage and noise current of an arbitrary resistor.*

2.4 The Origin of Noise

2.4.1 Resistor Thermal Noise

Probably the most well known noise source is the thermal noise of a resistor (also called Johnson noise). It is white noise since the PSD of the noise signal is flat throughout the frequency band. The noise is also called gaussian which means the amplitude of the noise signal has a gaussian distribution. The noise power is proportional to absolute temperature. The available noise power which is the same for every resistor is given by

$$P_{av,n} = kT\Delta f. \tag{2.54}$$

where k is Boltzmann's constant ($\sim 1.38 \times 10^{-23}$ J/K), T is the absolute temperature in K and Δf is the noise bandwidth in Hz. For ease of calculation, this available noise power is usually converted to a noise voltage source in series or a current source in parallel with the resistor as shown in Fig. 2.7. They are respectively given by

$$\overline{v_{nr}^2} = 4kTR\Delta f \qquad \text{and} \qquad \overline{i_{nr}^2} = \frac{4kT}{R}\Delta f. \tag{2.55}$$

Even though the available noise power is independent of the resistance, these voltage and current sources are not. Consequently the choice of a specific resistor is very important also from a noise point of view. For instance for a high impedance node, a minimum amount of injected noise current is desired. A high resistance is then preferred. However in series with an input voltage source a low resistance is preferred to keep the noise voltage low. The fact that they have the same available noise power is not relevant here.

2.4.2 Thermal Noise in MOS transistors

2.4.2.1 Classical MOS Channel Noise

It is quite clear that MOS transistors in the linear region need to display some sort of thermal noise. After all, the linear MOS transistor is essentially a controlled resistor. The drain noise current (Fig. 2.8) was calculated by [vdZ62]:

$$\overline{i_{nd}^2} = 4kT\gamma g_{d0}\Delta f, \tag{2.56}$$

where g_{d0} is the drain-source conductance at zero V_{DS}. Parameter γ is one at zero V_{DS} and — for long devices— decreases to a value of 2/3 in saturation. However, in short-channel NMOS

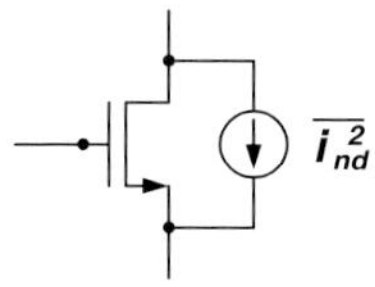

Figure 2.8: *Classical drain noise current for an NMOS transistor.*

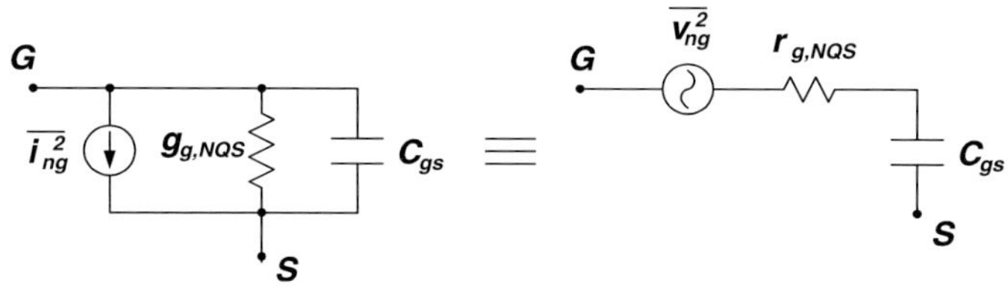

Figure 2.9: *Schematic of induced gate noise current and the equivalent voltage.*

devices the effective temperature of the carriers is significantly larger due to the high electric field in the channel. γ values of 2, 3 and more have been reported [Lee98]. Since the electric field for a fixed device is proportional to the V_{DS} it is important to keep this voltage as low as possible. Sometimes it may even be advised to use a non-minimum length transistor if this does not inhibit the required frequency performance. The PMOS transistor usually exhibits lower γ values than its NMOS counter part. Therefore, it could also be interesting to use a PMOS transistor for as far as other criteria allow this. In order to simplify calculations with MOS transistors in the saturation region, (2.56) is rewritten as

$$\overline{i^2_{nd}} = 4kT\frac{\gamma}{\alpha}g_m\Delta f, \tag{2.57}$$

where α is given by (2.42).

2.4.2.2 Induced Gate Noise

Since the gate is capacitively coupled with the channel, the drain noise also leads to a noisy gate voltage as shown by [vdZ86, Enz02].

$$\overline{i^2_{ng}} = 4kT\delta g_{g,NQS}\Delta f, \tag{2.58}$$

where

$$g_{g,NQS} = \omega^2 C^2_{gs} r_{g,NQS} \tag{2.59}$$

and δ is 4/3 for long-channel transistors as shown in [vdZ86]. This means that $\delta = 2 \times \gamma$. Since exceedingly hot carriers that increase the drain noise are also expected to increase the induced gate noise, it can be justified to state that also for short channels this equation remains valid. This was postulated in [Lee98].

The induced gate noise is clearly linked to the non-quasistatic gate resistance. In fact it can be considered as the thermal noise of this resistor. Consequently, the noise voltage $\overline{v_{ng}^2}$, (as shown in Fig. 2.9) may be expressed as

$$\overline{v_{ng}^2} = 4kT\delta r_{g,NQS}\Delta f, \tag{2.60}$$

Even though this expression is correct, care should be taken. Since the induced gate noise behaves partly as a capacitive reflection of the channel noise, both noise sources are not uncorrelated. The correlation coefficient for both noise currents, defined as

$$c \triangleq \frac{\overline{i_{ng} \cdot i_{nd}^*}}{\sqrt{\overline{i_{ng}^2} \cdot \overline{i_{nd}^2}}} \tag{2.61}$$

is equal to $j0.395$ for long-channel MOS transistors as shown by [vdZ86]. For ease of calculation it is assumed throughout the rest of this work that

$$c = j0.4, \tag{2.62}$$

for all regimes.

2.4.3 1/f Noise

Even though 1/f noise, pink noise or Flicker noise is very important in CMOS RF receivers, it will not be discussed in this text. As the name already stipulates, this type of noise has a PSD increasing towards low frequencies. Hence it will not be significant in low noise amplifiers operating in the GHz range. It will however have a prominent impact on the behavior and design of VCO's and down-conversion mixers. For the sake of completeness the most common expression for the PSD of the 1/f noise current in a MOS transistor is given below:

$$\overline{i_{nf}^2} = \frac{K_f}{f} \cdot \omega_T^2 \cdot WL \cdot \Delta f, \tag{2.63}$$

where K_f is a constant, f is the frequency, ω_T is the cut-off pulsation and WL is the transistor area. The location of the noise source is identical to that of the classical drain noise represented in Fig. 2.8.

2.4.4 Shot Noise

This noise mechanism was first introduced by Schottky and is based on the discrete nature of electrical charge. It occurs when a current flow crosses a potential barrier where the discreteness of the arrival times of the individual charges give rise to the noise current. Equivalent to thermal noise the PSD is flat and hence it is also a sort of white noise. The shot noise current is given by:

$$\overline{i_{nsh}^2} = 2qI_{DC}\Delta f, \tag{2.64}$$

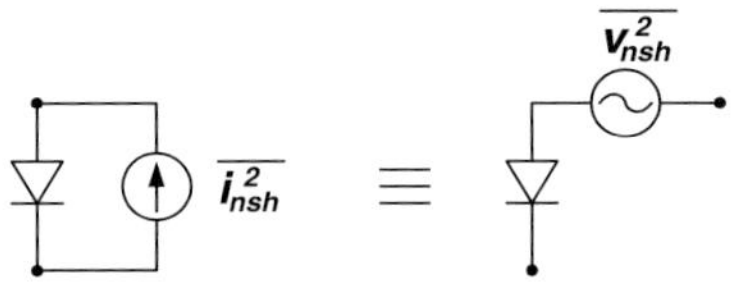

Figure 2.10: *Shot noise in silicon diodes.*

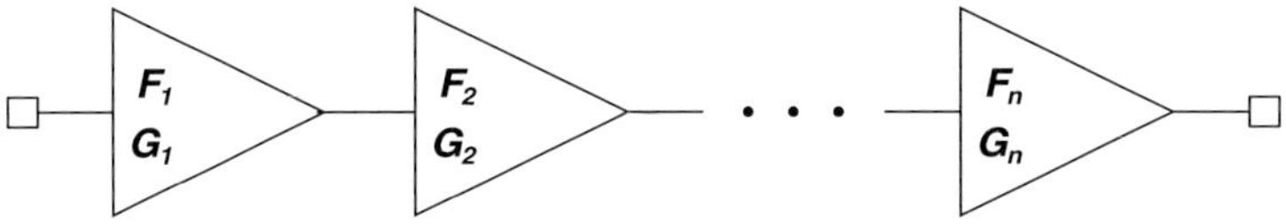

Figure 2.11: *Noise figure of a cascade of linear noisy gain blocks.*

where q is the elementary charge of an electron ($\sim 1.6 \times 10^{-19}\ C$) and I_{DC} is the DC current through the barrier.

In silicon, shot noise is most commonly found in pn-junctions (see Fig. 2.10). Consequently it is the most dominant noise source in bipolar transistors where large currents cross the base-collector junction.

2.5 The LNA in the Receiver Chain

2.5.1 Cascading Non-Ideal Building Blocks

2.5.1.1 Noise in a Cascade

Understanding what happens with the noise figure and distortion components when several non-ideal blocks are cascaded is crucial in the design of any receiver. Consider a cascade of linear gain blocks with power gain G_i and noise factor F_i as depicted in Fig. 2.11. Each block is assumed to be matched to 50 Ω at both input and output. The equivalent input noise of the final block —characterized by its noise factor F_n— can be referred to the input of the preceding block by dividing it by the gain of this block. This yields an equivalent noise factor for the cascade configuration of block n and block $n-1$ given by:

$$F_{n-1,n} = F_{n-1} + \frac{F_n - 1}{G_{n-1}}. \tag{2.65}$$

Continuing this technique all the way to the input of block 1 yields

$$F = F_1 + \sum_{i=2}^{n} \frac{F_i - 1}{\prod_{k=2}^{i} G_{i-1}}. \tag{2.66}$$

This means the noise added in each stage is suppressed by all the gains of the preceding stages. Hence the noise of subsequent stages becomes progressively less important. This can also be

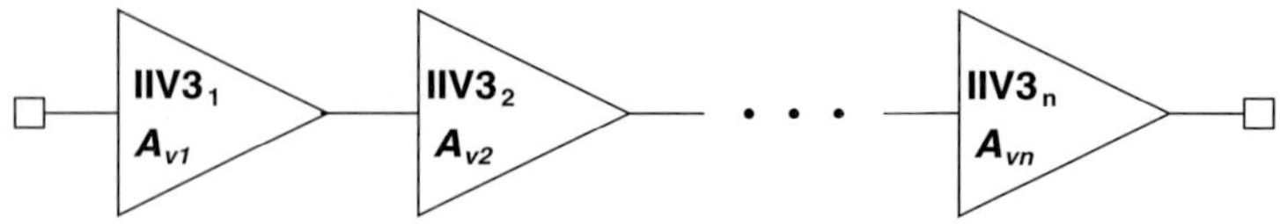

Figure 2.12: *IIP3 of a cascade of non-linear gain blocks.*

understood well from the fact that signal levels become higher proceeding through the cascade and additive noise becomes more and more negligible.

2.5.1.2 IIV3 of a Cascade

However as signal levels grow larger, the amount of distortion increases. Similar to the noise figure, also the IIV3 (the input referred intercept voltage) can be referred back to the input. Consider the cascade in Fig. 2.12. The blocks are now considered noiseless and non-linear and are characterized by their voltage gain A_{vi} and $\mathrm{IIV3}_i$. For simplicity the input impedance of each block is considered infinite while the output impedance is zero. Consider the output voltage of block 1. The third order intermodulation terms are

$$V_{im3,1} = \frac{A_{v1}V^3}{\mathrm{IIV3}_1^2} \tag{2.67}$$

where V is the input voltage amplitude. It is clearly seen that $V_{im3,1} = A_{v1}V$ for $V = \mathrm{IIV3}_1$ which follows directly from the definition of IIV3. The intermodulation terms at the output of block 2 consist of the amplified intermodulation terms at the output of block 1 and the intermodulation terms generated by block 2:

$$V_{im3,tot2} = A_{v2}V_{im3,1} + \frac{A_{v2}\left(A_{v1}V\right)^3}{\mathrm{IIV3}_2^2} \tag{2.68}$$

$$= \frac{A_{v2}A_{v1}V^3}{\mathrm{IIV3}_1^2} + \frac{A_{v2}\left(A_{v1}V\right)^3}{\mathrm{IIV3}_2^2}. \tag{2.69}$$

The total IIV3 of the cascaded blocks can now be calculated.

$$\frac{1}{\mathrm{IIV3}_{tot}^2} = \frac{V_{im3,tot2}}{A_{v2}A_{v1}V^3} \tag{2.70}$$

$$= \frac{1}{\mathrm{IIV3}_1^2} + \frac{A_{v1}^2}{\mathrm{IIV3}_2^2}. \tag{2.71}$$

This method can be extended for an arbitrary amount of blocks yielding the following formula for the total IIV3:

$$\frac{1}{\mathrm{IIV3}_{tot}^2} = \frac{1}{\mathrm{IIV3}_1^2} + \sum_{i=2}^{n} \frac{\prod_{k=1}^{i-1} A_{vk}^2}{\mathrm{IIV3}_i^2}. \tag{2.72}$$

This shows that the IIV3 of a block becomes more important as the amount of gain preceding the block increases, which means that the signals fed to this block grow larger. Since in a normal

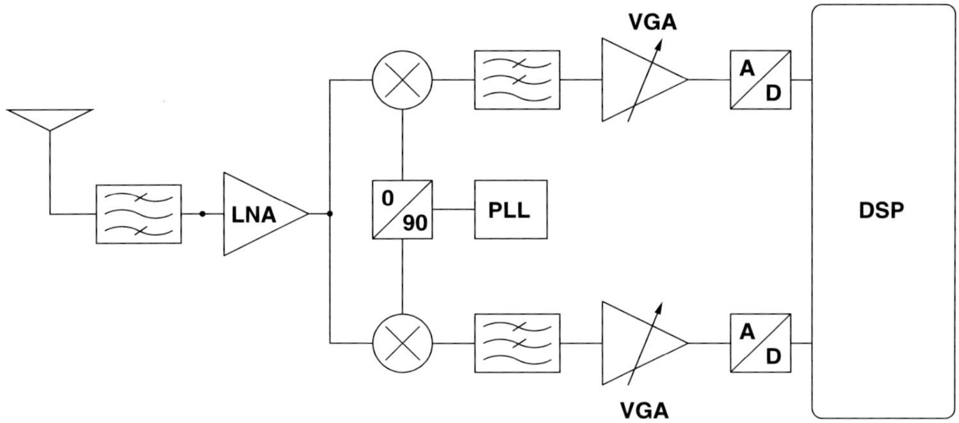

Figure 2.13: *Architecture of a direct conversion receiver.*

receiver the signals increase the further you proceed in the receiver chain, the IIV3 of the later blocks should be progressively higher than that of the first block.

2.5.2 Wireless Receiver Architectures

Clearly an ideal cascade of amplifying stages is no sufficient model for a wireless receiver. In general a receiver consists of an antenna, a band-select filter, a low-noise amplifier, one or more downconversion stages with a phase-locked loop and with or without interstage filtering, and an A/D converter with or without variable gain amplification. In classical superheterodyne receivers the downconversion was done in several steps. This required intermediate high-Q filtering for image cancellation after all stages. This topology is not interesting for a full CMOS wireless receiver since these high quality mixers have to be implemented externally and they would excessively increase the overall system cost. Therefore, two other topologies have been devised which don't need these high-Q filters: the zero-IF receiver [Abi95, Raz97] and the low-IF receiver [Cro98]. Both are direct conversion receivers since they don't require a second downconversion. The general schematic of this architecture is shown in Fig. 2.13.

In the zero-IF architecture, the LO has the same frequency as the RF-carrier. Consequently the RF-signal is its own image and hence cannot be filtered out as such. This is solved by mixing the RF signal with both sine and cosine of the LO. This is called quadrature mixing since a differential representation of both sine and cosine of the LO yields four 90 degrees shifted LO signals. If both signal paths are perfectly matched, the signals at the output of the mixers can be recombined (usually in the digital domain) to completely cancel the image, leaving only the wanted signal. The same technique is used for low-IF receivers. Since these topologies no longer require high-Q image filters, they are ideally suited for CMOS implementation. Moreover the image cancellation can be carried out in the digital back-end which is after all the core competence of CMOS.

The design of such direct-conversion receivers still poses some problems.

- Even though the actual image cancellation is done in the digital domain, it is the matching of the two quadrature paths in the analog front-end which determines the IMRR. Values of 30 to 40 dB have been achieved.
- Since there is no filtering in the RF front-end (the moderate band-select filtering aside), the signals reaching the ADC have a high dynamic range due to the possible presence of large blocking signals.
- Especially for zero-IF receivers, the presence of $1/f$ noise and DC-offset can severely limit the performance.

2.5.3 LNA Requirements

2.5.3.1 Matching

Referring to Section 2.2.3 the importance of the different types of matching for the input of the LNA will be discussed. Suppose the LNA is fed through a 50 Ω transmission line coming from the antenna or an off-chip band-select filter. First consider the power matching requirement. Since the MOS transistor is basically a voltage driven current source, an input power match is not required for a large output power. Moreover, it may be interesting to have an open circuit at the input since this would give the largest input voltage and hence the largest output current. Conclusion: power matching is not required.

However, for a 50 Ω source, the power match is identical to the impedance match. The reason for input impedance matching in the LNA is twofold. First, it avoids reflections over the transmission line feeding the LNA. And second it supplies a correct termination for the possible SAW-filter preceding the amplifier. This termination resistance is required in order to guarantee the frequency characteristic of the filter, both in the pass-band and in the stop-band. In pass-band, an incorrect termination resistance may lead to extra attenuation. The filter is the first block in the receiver. It attenuates the signal but due to its 50 Ω output impedance it has the same output noise power as the antenna (neglecting temperature differences). Hence an attenuation of 3 dB lowers the SNR with 3 dB which is equivalent to a noise figure of 3 dB. Any extra dB attenuation increases the noise figure with one dB. This must be avoided at all cost.

Also in the stop-band a correct termination is desired. Without it the attenuation in the stop-band could be reduced or the behavior of the stop-band ripple might be altered. Even though this has no direct influence on the wanted signal, it could lead to insufficiently suppressed blocker signals which can yield large in-band intermodulation products. Conclusion: a close to 50 Ω input impedance is very important in the signal band. It is also desired outside the signal band.

Noise matching aims at providing this equivalent source impedance to a given circuit which minimizes the noise figure of the circuit (cf. Appendix A). Often, the noise figure has quite a flat behavior around its optimum. such that an impedance match yields a sufficiently low noise figure. In classic microwave design, the amplifier (or transistor) is fixed once it has been chosen. The design is then done by choosing the equivalent source and load impedance that yield a

Specification	**Receiver 1**	**Receiver 2**
NF_{LNA}	3 dB	3 dB
$IIP3_{LNA}$	0 dBm	0 dBm
$A_{v,LNA}$	15 dB	30 dB
NF_{mix}	15 dB	15 dB
$IIP3_{mix}$	15 dBm	15 dBm

Table 2.2: *LNA and mixer specifications in Fig. 2.14.*

stable amplifier with a sufficient impedance match and a good noise figure. However in our case there are many more degrees of freedom. Consequently, the noise optimization can be done on transistor level while taking the impedance match as a constraint.

2.5.3.2 Noise Figure

Neglecting the channel-select filter, the LNA is the first building block in the receiver. As such it sets a lower bound on the attainable noise figure for the entire receiver. A very low noise figure is crucial. This becomes even more important for high sensitivity receivers like for the GPS system where the signal levels that need to be detected are extremely small.

2.5.3.3 Voltage Gain or Power Gain

The gain of the LNA should be large for more or less the same reason. It was learned from Section 2.5.1.1 that the noise of the stages following the LNA is suppressed by the gain of the LNA. Consequently for a receiver, the gain should be very large to minimize the noise figure contribution from the down-conversion mixer. Since the mixer is usually driven by a voltage, it is the voltage gain that should be optimized. Only if the LNA drives an external 50 Ω source (stand-alone LNA's), the power gain is considered in the optimization. For lab realizations and prototypes, the LNA output is often designed to drive 50 Ω as well, in order to ease the measurements.

In a real life environment however the LNA output stage is determined by the attached load, namely the input of the mixer. For a linear mixer this can in principal be either the capacitive load of the gate of a linear MOS transistor or the resistive load of its drain-source conductance. If the RF signal drives the gate, the voltage over the gate should be maximized and clearly the voltage gain is the main criterium. If the LNA drives the source of the mixing transistor, then the current through this transistor should be maximized. Since the current through the mixing transistor is proportional to the voltage over it, again the voltage gain should be maximized. For Gilbert type implementations, the load is always capacitive.

During the design of the LNA it might be tempting to just maximize the gain regardless of other building blocks or architectural considerations. However this is not advised. Increasing the gain of the LNA increases the signal levels in the mixer and this could give linearity problems. Equation (2.72) shows indeed that the IIP3 contribution of the mixer increases linearly with the gain of the LNA. For a Gilbert mixer which is not very linear, the gain of the LNA is usually kept

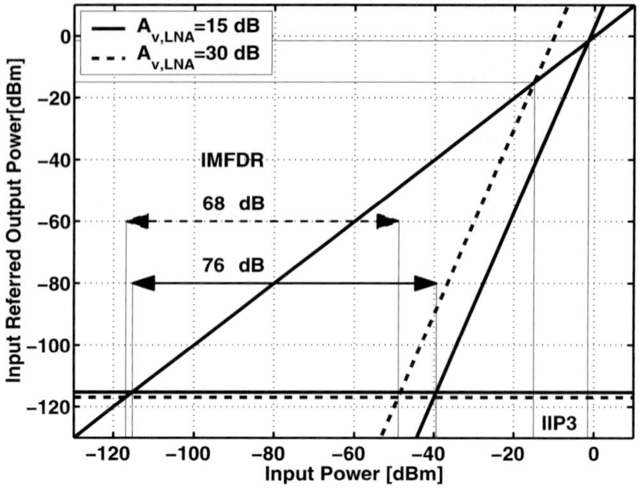

Figure 2.14: *Influence of the LNA gain on the IMFDR of a receiver.*

lower than when driving a linear mixer. To put it in another way, the voltage gain of the LNA should be set to maximize the dynamic range of the total receiver. If the next blocks are very linear but noisy, the gain is increased and vice versa. Fig. 2.14 shows the intermodulation free dynamic range (IMFDR) for a receiver consisting of an LNA and a mixer. IMFDR is defined by

$$\text{IMFDR [dB]} \triangleq \frac{2}{3}\text{IIP3 [dBm]} - \frac{2}{3}(P_{av,n} + P_{n,eq}) \text{ [dBm]}, \qquad (2.73)$$

where $P_{av,n} + P_{n,eq} = 10\log(kT) + \mathrm{NF}$ is the total input referred noise power of the receiver. It is often called the noise floor of the receiver since any signal below it is not visible on a spectrum analyzer. The IMFDR is the signal range between the level where the fundamental tone becomes visible and the level where the third order intermodulation terms become visible. The specifications of LNA and mixer used in Fig. 2.14 are listed in Table 2.2. The only difference between the two receivers is the LNA voltage gain of 15 dB and 30 dB respectively. Fig. 2.14 shows that the dynamic range of the receiver decreases with 8 dB when increasing the voltage gain of the LNA leaving the other specs unaltered. The reason is that the linearity of the mixer in this example is very (unrealistically) poor and the noise performance is rather good. Consequently the mixer is better off with lower signal powers.

It may be interesting to compare the high-level design of the LNA gain with the exposure control in a digital camera. The image sensor of the camera is representative for the fixed receiver front-end excluding the LNA. In high-end digital SLR cameras, a real-time histogram of the viewfinder image is used to set the correct exposure in order to fully exploit the dynamic range of the image sensor. If some region is too dark, the exposure needs to be increased, if it is too bright, the exposure can be reduced. A similar investigation of the possible signal levels allows to choose the best gain setting considering the dynamic range[1] of the subsequent building blocks The main difference is that the gain of the LNA is usually fixed after design. That is why, for the LNA, not one picture, but all possible pictures (read: signal levels) need to be considered in this gain optimization. In a complete receiver front-end design, of course, not only the gain of the LNA but all the specifications of the building blocks need to be considered in a global high-level optimization, also taking into account the power of blocking signals. This is done by investigating the minimum SNDR as the signal proceeds through the receiver.

2.5.3.4 Intermodulation Distortion

Similar to the receiver noise figure, which is lower bounded by the noise figure of the LNA, the IIP3 of the receiver is upper bounded by the IIP3 of the LNA. In many applications, the linearity specifications for the LNA don't pose many difficulties in the design of the receiver. Some applications have stronger linearity requirements, for instance because they need to be able to receive very large signals when the distance to the transmitter becomes small. Another reason can be the presence of large blocking signals in a neighboring band.

Even though the linearity requirements become more stringent further in the receiver path, the relative power of blocking signals usually decreases since each block has some intrinsic filtering. Consequently, the dynamic range of the signals is reduced further in the receiver, when the blocking signals are dominant in the linearity specification. This means the LNA requires the highest dynamic range of all receiver building blocks.

2.5.3.5 Reverse Isolation

The reverse isolation is defined as $-S_{12}$ where S_{12} is the reverse gain of the LNA. Basically three driving forces exist for increasing the reverse isolation. The first one is based on the spurious emission specification of the receiver. The signal coming from the local oscillator may couple back through the mixer to the output of the LNA. This signal can reach the antenna through the reverse gain of the LNA. The higher the reverse isolation, the smaller the spurious LO tone at the antenna. The amount of reverse isolation required depends on the LO signal amplitude, the coupling through the mixer and the allowed spurious signal level at the antenna. Usually a value of 25 dB to 30 dB is sufficient.

Another reason for increasing the reverse isolation is that the input matching becomes considerably more reliable when the reverse isolation is high. Often a low reverse isolation goes hand in hand with a reduced gain since the inherent feedback of the reverse gain reduces the signal efficiency.

A final driving force is the intrinsic stability of the amplifier which is discussed next.

---

[1]The dynamic range implies in fact two numbers here, the minimum and maximum signal level, not just the difference between them. Otherwise no conclusion about the required gain could be drawn.

2.5.3.6 Stability

Several techniques exist to describe and design stable amplifiers. For RF and microwave amplifiers, it is interesting to define the concept of unconditional stability. A circuit is unconditionally stable if for any combination of source and load impedance, the circuit is stable. A single parameter μ_s was defined in [Edw92][2] which can describe the unconditional stability of an amplifier as a function of its S-parameters:

$$\mu_s = \frac{1 - |S_{11}|^2}{|S_{22} - S_{11}^* \Delta| + |S_{21} S_{12}|} > 1 \tag{2.74}$$

where Δ is the determinant of the S-parameter matrix:

$$\Delta = S_{11} S_{22} - S_{12} S_{21}. \tag{2.75}$$

A completely equivalent but not equal parameter can be found by interchanging indexes 1 and 2. Parameter μ can be calculated by the simulator in order to investigate the stability of the amplifier. Generally μ increases for decreasing $|S_{12}|$. Hence, increasing the reverse isolation will ease the design for stability. Moreover, if $S_{12} = 0$ then (2.74) reduces to

$$\mu_s = \frac{1}{|S_{22}|} > 1 \qquad \text{or} \qquad |S_{22}| < 1. \tag{2.76}$$

Again indexes 1 and 2 can be interchanged.

For integrated CMOS LNA's, simulating or designing for unconditional stability is quite difficult, especially at high frequency. This is due to the lack of RF-models for the devices as they are laid out by the designer. Some companies provide S-parameter models for a limited set of devices which can be implemented with parameterized cells. Simulating the stability with those cells is possible and yields good results when the S-parameters have been extracted correctly from test chip measurements. Drawback of the use of these devices is that it takes away a lot of the design freedom.

2.5.3.7 Single-ended vs. Differential

The main reason for using a differential topology is the common mode rejection. This has some major advantages. The noise on the power supply lines appears as common-mode and is not seen at the output. This largely reduces possible problems with digital switching noise when integrating both RF, baseband analog and digital on the same die. Moreover, the even harmonics also appear as common mode and are similarly suppressed. Some blocking signals that have a second harmonic in the signal band are this way largely disarmed.

The main drawback of differential topologies is that they require more or less twice the power for the same performance. In handheld or wearable applications, where power consumption is the bottleneck, this is often unacceptable. Naturally this trade-off should be considered for every application separately.

[2] The original symbol is μ but μ_s is used here to avoid confusion with the mobility μ.

2.6 Topologies for Low-Noise Amplifiers

2.6.1 The Inductively Degenerated Common Source LNA

2.6.1.1 From Basic Common-Source Amplifier to Inductively Degenerated Common-Source LNA

This section will discuss the origin and the basic schematic of one of the most popular LNA topologies, known as the inductively degenerated common source LNA. Based on Fig. 2.15, the gradual evolution from basic common source amplifier to inductively degenerated CS LNA is explained. The circuit, depicted in Fig. 2.15(a) shows a simple baseband one-transistor CS amplifier. The first LNA criterium is already fulfilled, the positive gain requirement. One of the problems with this circuit is that it has a purely capacitive input impedance (at least according to the classical quasi-static MOS model).

In order to create a resistive input, it suffices to place a termination resistor parallel to the LNA input (Fig. 2.15(b)). In this figure, the termination resistor is connected to ground. In reality this would upset the DC biasing of the amplifier. Consequently, resistor R_t should be connected to the DC biasing node. This node can be decoupled from the ground with a large decoupling capacitor $C_{dc,in}$. Therefore, R_t is connected to AC ground and the AC performance ($\omega \gg 1/(R_t C_{dc,in})$) remains unaltered. The input impedance for $\omega \gg 1/(R_t C_{dc,in})$ is given by

$$Z_{in} = \frac{R_S}{1 + \frac{j\omega}{\omega_p}}, \tag{2.77}$$

where

$$\omega_p = \frac{1}{R_S(C_{gs} + MC_{gd})}, \tag{2.78}$$

M is the Miller factor and $R_S = 50\ \Omega$, the source resistance. The Miller effect for this circuit is very pronounced,

$$M = 1 + g_{m1}R_L. \tag{2.79}$$

This effect strongly limits the frequency performance and gives rise to a very poor reverse isolation.

Adding a cascode transistor as shown in Fig. 2.15(c) significantly decreases the Miller effect since it is now decoupled from the gain of the circuit. If $R_L \ll r_{ds2}$ where r_{ds2} is the output resistance of cascode transistor M2, then

$$M = 1 + \frac{g_{m1}}{g_{m2}} \approx 2. \tag{2.80}$$

One of the problems with this circuit is that R_L needs to be large for a high gain. However this will cause a large DC voltage drop over R_L. In order for the circuit to operate within parameters, the voltages over M1 and M2 need to be larger then $V_{DS,sat} = V_{GS} - V_T$. Therefore,

$$R_L < \frac{V_{DD} - V_{DS,sat,1} - V_{DS,sat,2}}{I_{DC}}, \tag{2.81}$$

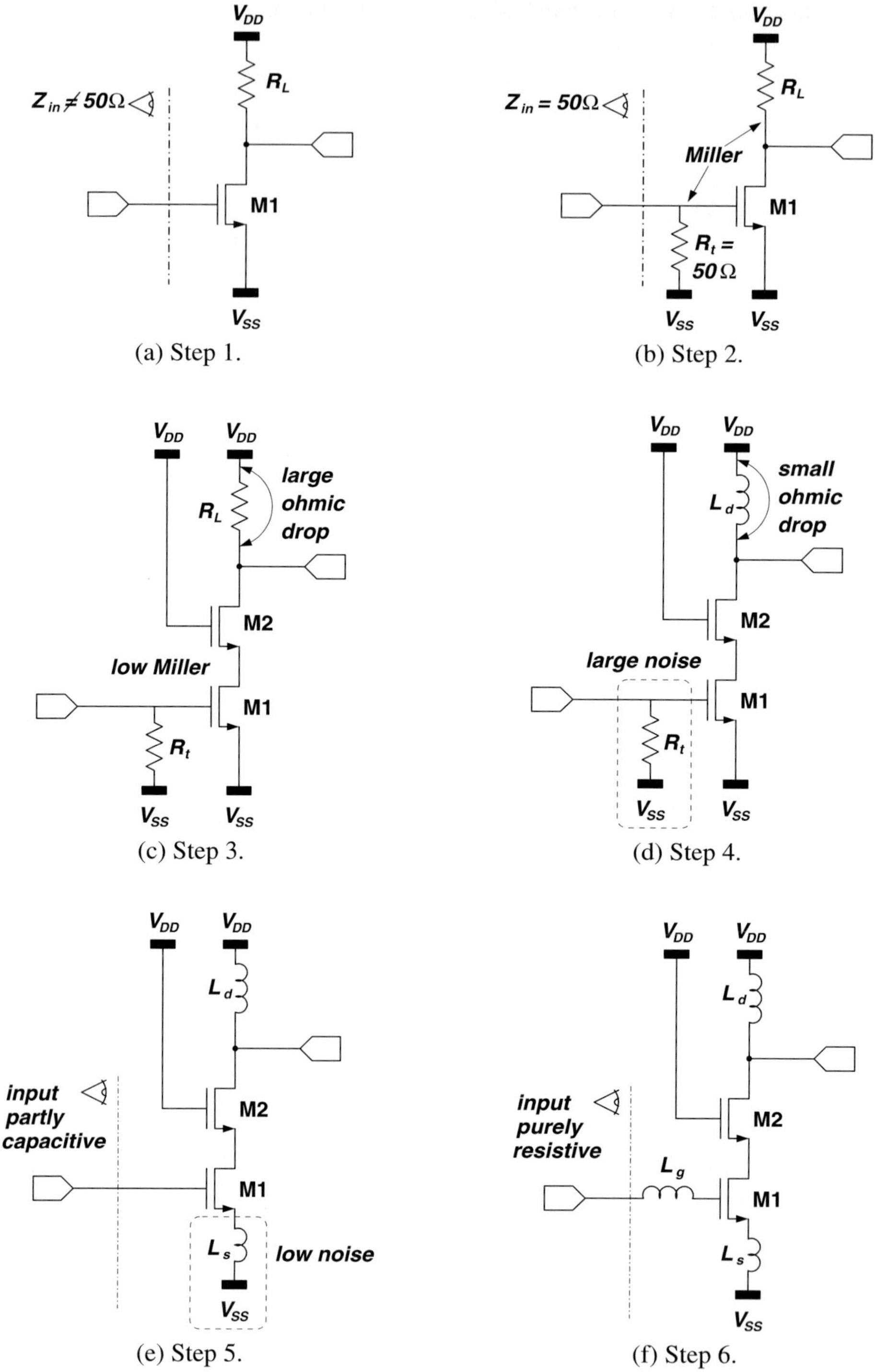

(a) Step 1. (b) Step 2. (c) Step 3. (d) Step 4. (e) Step 5. (f) Step 6.

Figure 2.15: *Gradual evolution from basic CS amplifier to inductively degenerated CS LNA.*

which is usually limited to a few hundred Ohm. Suppose both $V_{DS,sat}$ are 0.25 V and V_{DD} is 1.5 V. The voltage gain of the circuit is then limited by

$$A_v < g_{m1} R_L = 2 \times \frac{V_{DD} - V_{DS,sat,1} - V_{DS,sat,2}}{V_{DS,sat,1}} = 8, \tag{2.82}$$

which corresponds to 18 dB.

One option to alleviate this is by using an active load element. If R_L is replaced by a PMOS transistor then the voltage drop over the load transistor can be limited by its own saturation voltage, which is a lot lower than the ohmic drop over a resistor with the same impedance. Disadvantages are the extra output capacitance, the need for an extra biasing node, the limited output swing, the fact that the gain is dependent on the r_{ds} of a transistor which is not accurately modelled and the non-linear behavior of the load transistor. The most important disadvantage is the noise introduced by the PMOS. The squared noise current of a load resistance is inversely proportional to its resistance,

$$\overline{i^2_{n,R_L}} = \frac{4kT\Delta f}{R_L}, \tag{2.83}$$

but the squared noise current of a load transistor M_p is proportional to its transconductance g_{mp},

$$\overline{i^2_{n,M_p}} = 4kT\gamma g_{mp} \Delta f, \tag{2.84}$$

which can be larger with a factor of 20 or more. This is unacceptable for a low noise amplifier.

Luckily another option is available for LNA's in wireless receivers. The LNA should only amplify in a relatively small frequency band around the carrier. Therefore, it is possible to replace R_L with an inductor L_d (see Fig. 2.15(d)). This inductor should go into parallel resonance with the parasitic capacitance on the output node precisely at the carrier frequency f_0. If the quality factor Q_L of the parallel tank was infinite, the equivalent load impedance and the gain at f_0 would be infinite. However due to the losses in the tank, the equivalent load resistance remains finite. Considering only the loss in the series resistance of the load inductor, the equivalent load resistance is given by

$$R_L = R_{L,s}(Q_L^2 + 1), \tag{2.85}$$

where $R_{L,s}$ is the series resistance of the load inductor and

$$Q_L = \frac{\omega_0 L_d}{R_{L,s}}. \tag{2.86}$$

Now, the DC voltage drop over the load is proportional to the DC series resistance of the inductor $R_{L,s}$ which is very small, in the order of a few Ohm. The gain is proportional to R_L which can be made large: a few hundred to over a thousand Ohm. The squared noise current from the load is inversely proportional to R_L. The noise factor of this circuit, neglecting the contribution of M2 is approximately

$$F \approx 2 + \frac{\gamma}{\alpha} \frac{4}{g_{m1} R_S} + \frac{4}{g_{m1}^2 R_S R_L}, \tag{2.87}$$

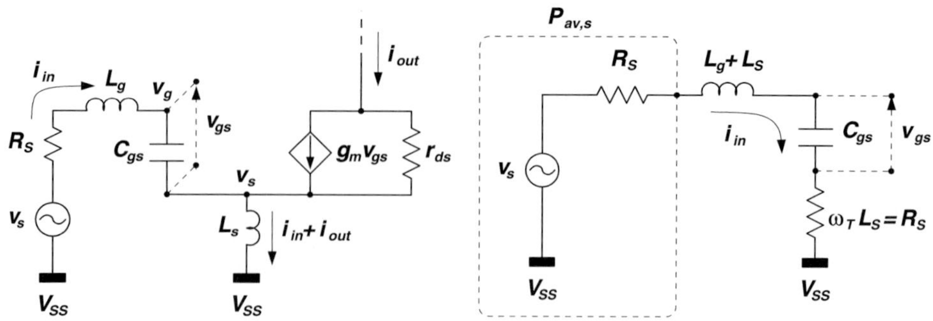

(a) Small-signal input schematic. (b) Equivalent input schematic.

Figure 2.16: *The origin of the resistive component in the input impedance of the CS LNA and the input quality factor* Q_{in}.

Real noise figure values for this type of amplifiers are in the order of 6 dB or more. This is too large for most applications. Another type of input matching is required.

Instead of adding a termination resistor, an inductor L_s is connected between the source of M1 and V_{SS} as shown in Fig. 2.15(e). Maybe somewhat unexpectedly, this creates a resistive part in the input impedance. Neglecting the gate-drain capacitance, the input impedance is given by

$$Z_{in} = \frac{1}{j\omega C_{gs}} + j\omega L_s + \omega_T L_s, \tag{2.88}$$

where the cut-off pulsation ω_T is defined by (2.40). At normal operating frequencies, the input still behaves capacitively due to C_{gs}. Therefore an extra gate resonance inductor L_g is connected in series with the input as indicated in Fig. 2.15(f). The input impedance is now

$$Z_{in} = \frac{1}{j\omega C_{gs}} + j\omega(L_g + L_s) + \omega_T L_s, \tag{2.89}$$

which reduces to

$$Z_{in} = \omega_T L_s, \tag{2.90}$$

at the operating frequency f_0 when

$$\omega_0(L_g + L_s) = \frac{1}{\omega_0 C_{gs}}. \tag{2.91}$$

In the design, L_s is chosen such that $\Re(Z_{in}) = \omega_T L_s = 50\ \Omega$. L_g is chosen according to (2.91) so that $\Im(Z_{in}) = 0$ at ω_0.

In what follows the subscript 1 for parameters relating to M1 will be left out for reasons of simplicity.

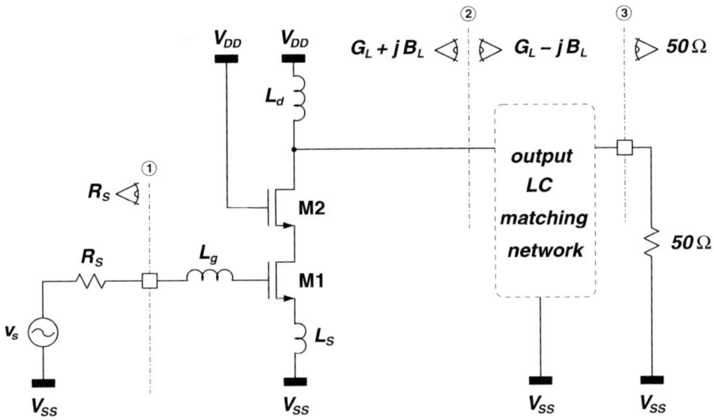

Figure 2.17: *Simplified schematic of the CS LNA matched at both input and output.*

2.6.1.2 Power Gain

Before calculating the power gain it is interesting to see where the resistive part of the input impedance comes from. Suppose a certain current i_{in} is flowing through the C_{gs} of M1 as depicted in Fig. 2.16(a). This will cause a v_{gs} which is lagging 90 degrees with respect to i_{in}. Hence, a current i_{out} will flow given by $g_m v_{gs}$ which is in phase with v_{gs} and also lagging 90 degrees with respect to i_{in}. This current (together with i_{in} itself) will flow through L_s giving rise to a voltage over L_s leading 90 degrees with respect to i_{out} and in phase with i_{in}:

$$\begin{aligned} v_s &= j\omega L_s(i_{in} + i_{out}) \\ &= i_{in}(j\omega L_s + \frac{g_m}{C_{gs}} L_s). \end{aligned} \tag{2.92}$$

Since v_s is part of v_{in} according to

$$v_{in} = j\omega L_g i_{in} + v_{gs} + v_s, \tag{2.93}$$

this in phase component is found also in Z_{in} given in (2.89).

For the derivation of the power gain it is assumed that both the input and output of the LNA are matched to 50 Ω. For the output this can be done with any lossless matching network as depicted in Fig. 2.17. Specific matching networks will be discussed in more detail in Section 4.8. The small signal equivalent of Fig. 2.17 is shown in Fig. 2.18. The available source power $P_{av,s}$ is given by

$$P_{av,s} = \frac{v_s^2}{4R_S}. \tag{2.94}$$

Since the input is matched, the input current i_{in} is

$$i_{in} = \frac{v_s}{2R_S} = \sqrt{\frac{P_{av,s}}{R_S}} \tag{2.95}$$

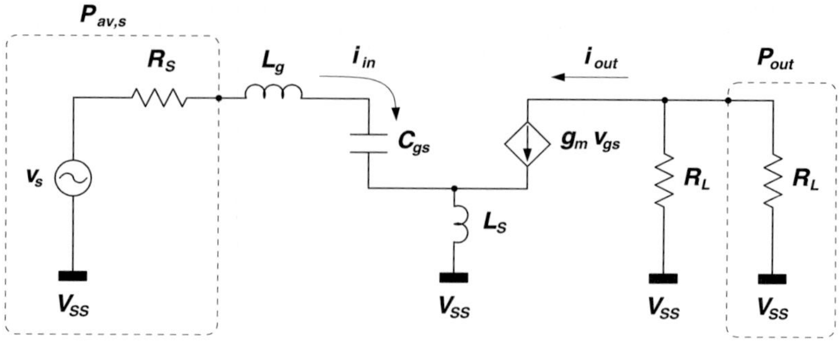

Figure 2.18: *Simplified small signal schematic of the LNA in Fig. 2.17.*

The input current gives rise to an output current according to

$$i_{out} = i_{in}\frac{\omega_T}{\omega_0}, \tag{2.96}$$

which is equivalent to the current gain β of a bipolar transistor. Since the output of the circuit is matched it is also matched at reference plane ② in Fig. 2.17. Both to the left and right of the reference plane, the current i_{out} sees a resistance R_L (also shown in Fig. 2.18). Thus only half of the output current is used to generate the output power. Consequently

$$P_{out} = \left(\frac{i_{out}}{2}\right)^2 R_L. \tag{2.97}$$

Equations (2.94) to (2.97) allow calculation of the power gain:

$$G_T = \frac{P_{out}}{P_{av,s}} = \frac{R_L}{4R_S}\left(\frac{\omega_T}{\omega_0}\right)^2. \tag{2.98}$$

It is clear that the power gain increases with the load resistance. However practically the load resistance achievable with an on-chip inductor is limited as will be shown in Section 4.10.2. Notice also that the power gain goes up with increasing ω_T. As such, deeper submicron technologies automatically improve the gain of the LNA. Fig. 2.19(b) illustrates the behavior of the power gain in the M1 design space assuming a fixed load resistance of 500 Ω. This assumption is not entirely justified since a smaller M1 will usually result in a smaller output capacitance. This will allow a larger load inductor and hence a larger load resistance as will be shown in Section 4.8. For now, this is neglected. It is clearly seen in Fig. 2.19(b) that the gain is only dependent on $V_{GS} - V_T$ through ω_T.

We can rewrite (2.98) as

$$G_T = \frac{R_L}{4R_S}\left(\frac{\omega_T}{\omega_0}\right)^2 = Q_{in}^2 g_m^2 R_L R_S, \tag{2.99}$$

where

$$Q_{in} = \frac{v_{gs}}{v_s} = \frac{1}{2\omega_0 C_{gs} R_S}, \tag{2.100}$$

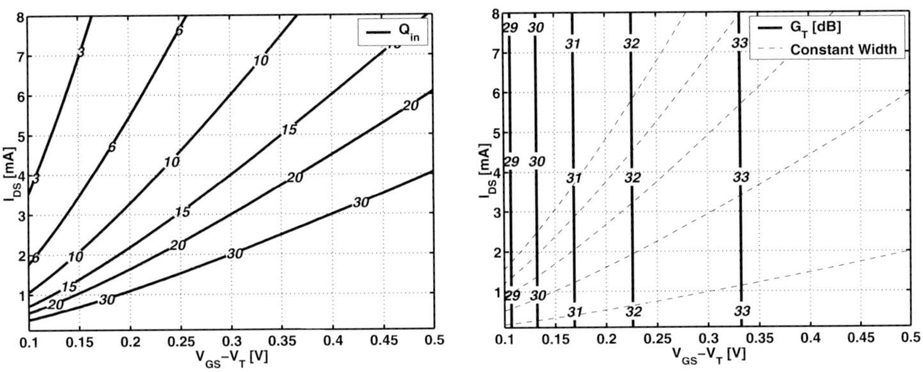

(a) Q_{in}[ ]. (b) G_T [dB].

Figure 2.19: *Contours of the input quality factor and power gain of the CS LNA.*

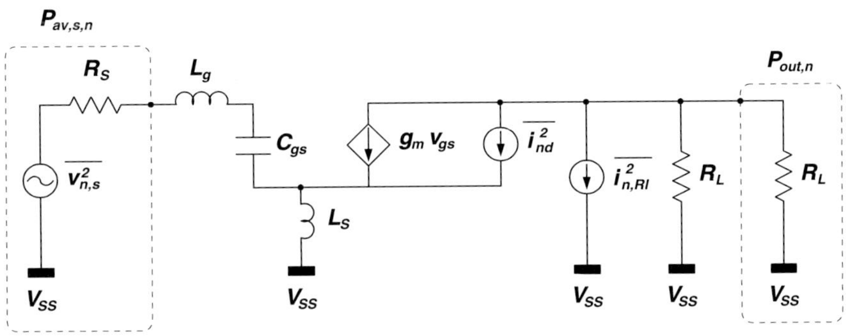

Figure 2.20: *Simplified small signal schematic of the CS LNA in Fig. 2.17 with noise sources.*

represents the quality factor of the series resonant input tank consisting of $2R_S$, $(L_g + L_s)$ and C_{gs} as indicated in Fig. 2.16(b). Contours of Q_{in} are shown in Fig. 2.19(a). This explains why G_T is only a function of $V_{GS} - V_T$. Equation (2.99) shows that the gain increases with g_m, which comes to no surprise. However, if g_m is increased for a constant $V_{GS} - V_T$ by increasing the width of M1, the gain remains constant. The reason is that the larger width and hence the larger C_{gs}, causes a proportional reduction in the quality factor of the input tank. This completely counters the increased g_m.

2.6.1.3 Noise Figure

For calculation of the noise factor of the LNA, only the classical channel noise of M1 and the thermal noise of R_L are considered as indicated in Fig. 2.20. The noise contribution of M2 and other parasitic noise sources are ignored. The derivation was based on the quasi static approximation and the influence of the cascode pole, the Miller-effect and other parasitics are neglected.

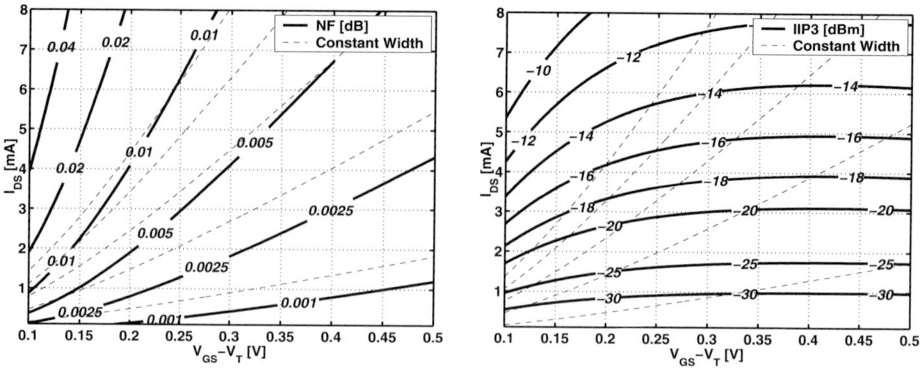

(a) NF [dB]. (b) IIP3 [dBm].

Figure 2.21: *Contours of the noise figure and IIP3 of the CS LNA.*

A more in depth discussion is given in Chapter 4 where also the influence of the main parasitics is covered. For now, the following expression approximates the noise factor of the amplifier:

$$\begin{aligned} F &\approx 1 + (F_d - 1) + (F_L - 1) \\ &\approx 1 + \frac{\gamma}{\alpha} g_m R_S \left(\frac{\omega_0}{\omega_T}\right)^2 + 4\left(\frac{\omega_0}{\omega_T}\right)^2 \frac{R_S}{R_L}, \end{aligned} \tag{2.101}$$

where $(F_d - 1)$ and $(F_L - 1)$ denote the contributions of transistor M1 and the equivalent load resistor R_L respectively. It might be surprising to find g_m in the numerator of the noise factor rather than in the denominator. However, $F_d - 1$ can be rewritten as

$$F_d - 1 = \frac{\gamma}{4\alpha g_m R_S} \cdot \frac{1}{Q_{in}^2}, \tag{2.102}$$

where Q_{in} is given by (2.100) and accounts for the input series resonance. In this equation g_m is found in the denominator as intuitively expected. This proportionality only holds for a fixed device width since then Q_{in} is a constant. The factor 4 in the denominator finds its origin in the fact that only half of the drain noise current finds its way to the output owing to the feedback inductor L_s. Note also that $(F_L - 1) = G_T^{-1}$ goes down with increasing R_L since an infinite R_L gives a zero noise current at the output. It also decreases with increasing $\frac{\omega_T}{\omega_0}$ since the gain of the circuit increases and the noise contributed by the load resistor becomes proportionally less important.

It is even more interesting to take a closer look at the behavior of $(F_d - 1)$ with regards to the dimensions of M1:

$$(F_d - 1) \approx \frac{\gamma}{\alpha} g_m R_S \left(\frac{\omega_0}{\omega_T}\right)^2 \propto \frac{g_m}{\omega_T^2}. \tag{2.103}$$

In order to illustrate the behavior of F_d a contour plot is depicted in Fig. 2.21(a). It is based on hand calculations with parameters extracted from the 0.25 μm Kawasaki technology in Table 2.1 and for an operating frequency of 1.57 GHz conform the primary GPS-band.

- For a fixed current through the device, (2.103) becomes

$$(F_d - 1) \propto \frac{1}{(V_{GS} - V_T)^3} \propto W^{3/2}. \tag{2.104}$$

  This shows that the noise figure decreases drastically with increasing $V_{GS} - V_T$, even if the current through the device does not change. This can be deduced also from Fig. 2.21(a) since a constant current is simply a horizontal line. Going to the right with increasing $V_{GS} - V_T$ yields a gradually lower noise figure.

- For a fixed device width,

$$(F_d - 1) \propto \frac{1}{(V_{GS} - V_T)} \propto \frac{1}{\sqrt{I}}. \tag{2.105}$$

  This means that for a given device if the bias voltage is increased, the noise figure improves due to the increase in ω_T^2 which is faster than the increase in g_m. This can easily be tested experimentally since after processing the device width tends to remain fixed. Contours for constant width are also represented in Fig. 2.21(a). Increasing either I_{DS} or $V_{GS} - V_T$ (following the contours from left to right) indeed shows a slowly decreasing noise figure.

- For a fixed overdrive voltage $V_{GS} - V_T$, the noise factor behaves as

$$(F_d - 1) \propto W \propto I. \tag{2.106}$$

  This is undoubtedly the strangest behavior; the noise figure increases with increasing current! It is illustrated graphically in Fig. 2.21(a) where the contours for constant $V_{GS} - V_T$ are just vertical lines. Increasing the current (following the vertical lines upward) indeed gives an increasing noise figure. This is very counterintuitive but can be understood as follows. A fixed $V_{GS} - V_T$ implies a fixed ω_T which means the current gain from (2.96) is also fixed and the squared output noise current due to the source is unchanged. However the squared output noise current from M1 is proportional to its g_m and hence proportional to the bias current for a fixed $V_{GS} - V_T$. Therefore the noise figure increases with increasing bias current.

As well $(F_L - 1)$ as $(F_d - 1)$ decrease with increasing ω_T/ω_0 which means that smaller technologies also improve the inherent classical noise contribution in the noise figure. However, the excess noise factor γ tends to increase for smaller gate lengths as mentioned in Section 2.4.2. The reason is that the electric field within the short channel increases causing an electron temperature notably larger than the lattice temperature. To decrease γ it is advised to lower the drain-source voltage in order to keep the electric field in the channel as low as possible. Increasing the gate length to reduce γ makes no sense since the increase in ω_T would far outweigh the benefits of a lower γ.

A final possibility for improvement appears to be given by the decrease of R_S (which would also increase the power gain as seen in (2.98). Of course, the actual source resistance is the output resistance of the antenna or the bandpass filter preceding the LNA and is consequently fixed. However it is possible to use a lossless transformation network to synthesize an equivalent source resistance (analogous to the matching network at the output in Fig. 2.17) . Actually such a transformation network is inherently present albeit that it normally increases the equivalent source resistance. This will yield an even larger contribution of the classical channel noise in the noise figure (but it will reduce the contribution of the non-quasistatic induced gate noise). These issues will be clarified in Chapter 4.

2.6.1.4 Linearity

In Section 2.3.2 it was stated that for any NMOS in saturation, the third order intermodulated output current is equal to the fundamental output current ($\mathrm{IM3} = 1$) for an input voltage amplitude given by (2.47). This goes also for transistor M1 in Fig. 2.17. Consequently, referring to the small-signal equivalent in Fig. 2.18, the v_{gs} at which $\mathrm{IM3} = 1$ is given by (2.47):

$$\mathrm{IV3}\ [\mathrm{V\ amp}] = \sqrt{\frac{4}{3}\frac{V_{GST}\left(2+\Theta V_{GST}\right)\left(1+\Theta V_{GST}\right)^2}{\Theta}}. \tag{2.107}$$

Equations (2.94) and (2.100) allow calculation of the available source power corresponding with this v_{gs}

$$\mathrm{IIP3}\ [\mathrm{W}] = \frac{\mathrm{IV3}^2}{2\times 4Q_{in}^2 R_S} = \frac{\mathrm{IV3}^2\omega_0^2 C_{gs}^2 R_S}{2}. \tag{2.108}$$

Together with (2.107) this results in

$$\mathrm{IIP3}\ [\mathrm{W}] = \frac{1}{8R_S}\frac{4}{3}\frac{V_{GST}\left(2+\Theta V_{GST}\right)\left(1+\Theta V_{GST}\right)^2}{\Theta}\frac{1}{Q_{in}^2}, \tag{2.109}$$

the available source power expressed in Watt. Converting this expression to [dBm] yields,

$$\mathrm{IIP3}\ [\mathrm{dBm}] = 5.25 + 10\log\left(\frac{V_{GST}\left(2+\Theta V_{GST}\right)\left(1+\Theta V_{GST}\right)^2}{\Theta}\right) - 20\log\left(Q_{in}\right). \tag{2.110}$$

Substituting (2.100) in (2.110) gives

$$\mathrm{IIP3} = 5.25 + 10\log\left(\frac{V_{GST}\left(2+\Theta V_{GST}\right)\left(1+\Theta V_{GST}\right)^2}{\Theta}\right) + 20\log\left(2\omega_0 R_S C_{gs}\right), \tag{2.111}$$

The first term in this equation is fixed and consists of the product of the conversion factor for going from V to mW $(10\log(1000/(2\times 50))$ and the factor 4/3 in the IV3. The second term represents the IV3 of M1 itself while the last term is the quality factor of the input stage. This is logical indeed, if Q_{in} is higher then v_{gs} will be larger for the same input power, thus IIP3 decreases. However it should be noted that the output referred 3rd order intermodulation intercept point (OIP3) is independent of Q_{in}. It is given by

$$\mathrm{OIP3} = \mathrm{IIP3}\times G_T. \tag{2.112}$$

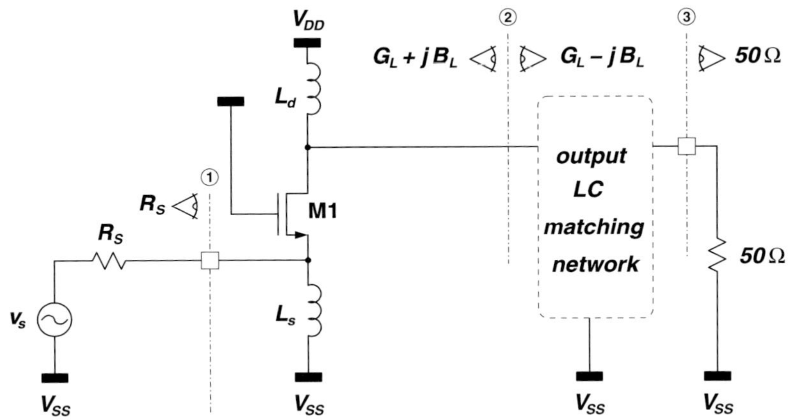

Figure 2.22: *Simplified schematic of the CG LNA matched at both input and output.*

substituting G_T by (2.99) results in

$$\mathrm{OIP3} = 5.25 + 10\log\left(\frac{V_{GST}\left(2+\Theta V_{GST}\right)\left(1+\Theta V_{GST}\right)^2}{\Theta}\right) + 10\log\left(g_m^2 R_L R_S\right), \quad (2.113)$$

independent of Q_{in}.

Fig. 2.21(b) plots the contours of IIP3 in the design space of M1. It is seen that the IIP3 is quite a weak function of $V_{GS} - V_T$ where for a simple NMOS it is only dependent on $V_{GS} - V_T$. The reason is the third term in (2.110) representing the input quality factor, defined by (2.100). Q_{in} is inversely proportional to the width of M1:

$$Q_{in} = \frac{1}{2\omega_0 C_{gs} R_S} \propto \frac{1}{W}. \quad (2.114)$$

Hence, when increasing $V_{GS} - V_T$, at a fixed current, the device width is inversely proportional to the square root of the overdrive voltage. Consequently the second term of (2.110) increases but the third term decreases. These effects almost cancel each other resulting in the relatively flat behavior of the contours. The only sure way to improve the linearity of the LNA is to increase the current consumption.

2.6.2 The Common-Gate LNA

The common-gate LNA, depicted schematically in Fig. 2.22 is the main competitor of the inductively degenerated common-source LNA. The reason is quite obviously the resistive input impedance. In the common-gate amplifier the signal is fed to the source of the input transistor rather than the gate. The input resistance is the inverse of the transconductance. Simply equating it to 50 Ω or 20 mS results in a correct impedance match. Actually, things are a bit more complicated but this basic principle is the reason for its popularity.

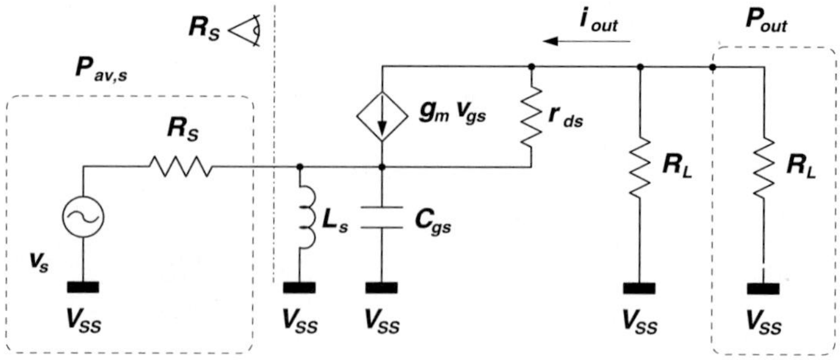

Figure 2.23: *Simplified small signal schematic of the CG LNA in Fig. 2.22.*

2.6.2.1 Input Matching

The small signal equivalent of the CG LNA is shown in Fig. 2.23. The following discussion is based on the classical MOS-model as described in Section 2.3.1. No non-quasistatic effects are taken into account at this point. The impedance seen at the input of the LNA (neglecting the input capacitance) is then

$$Z_{in} = \frac{1}{g_m + g_{mb}} \frac{2r_{ds} + R_L}{2r_{ds}}. \tag{2.115}$$

It can be set to equal the source resistance, R_S. For larger g_m, the required load resistance for acquiring a correct input impedance increases rapidly. Once M1 is fixed in its design space, the value of R_L for a 50 Ω input impedance can be found as:

$$R_L = 2\left(R_S\left(g_m + g_{mb}\right) - 1\right) r_{ds} = 2(n g_m R_S - 1) r_{ds}, \tag{2.116}$$

where

$$n = \frac{g_m + g_{mb}}{g_m}. \tag{2.117}$$

Substituting (2.41) in (2.116) yields

$$R_L = 2\frac{n g_m R_S - 1}{\Lambda I_{DS}} = \frac{4 n R_S}{\Lambda\left(V_{gs} - V_T\right)} - \frac{2}{\Lambda I_{DS}}. \tag{2.118}$$

If the value of R_L, calculated from this expression becomes negative, an input match can no longer be realized without an extra matching network. Therefore it is required that

$$g_m > \frac{1}{n R_S} \approx 17 \text{ mS}. \tag{2.119}$$

A serious disadvantage of the input impedance of the CG LNA is that it depends heavily on the value of r_{ds}. This resistance is not well known and not accurately modelled. An offset of 50% is quite possible. Consequently also the input impedance will show a large spread on process variations. This can be avoided by placing an extra resistor R_{dsx} between drain and

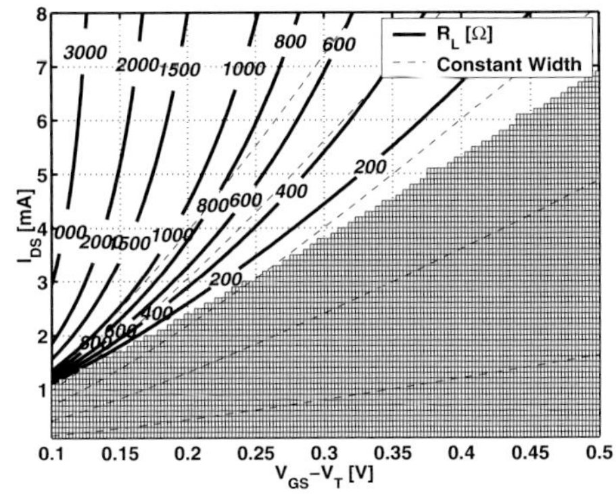

(a) The required load resistance R_L [Ω].

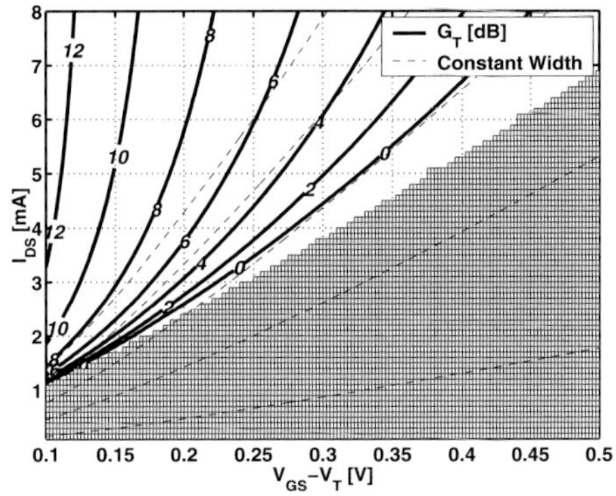

(b) G_T [dB].

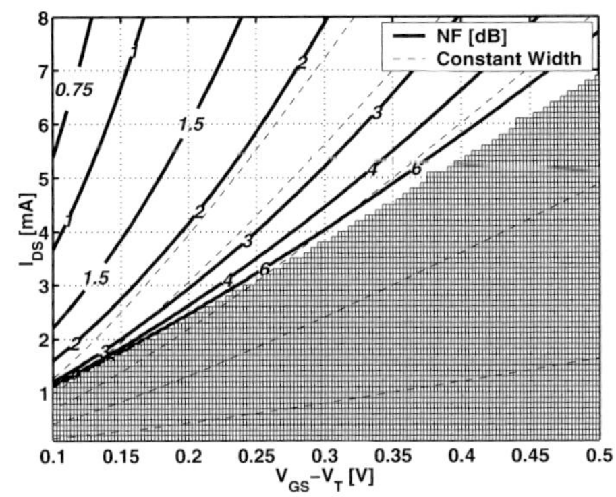

(c) NF [dB].

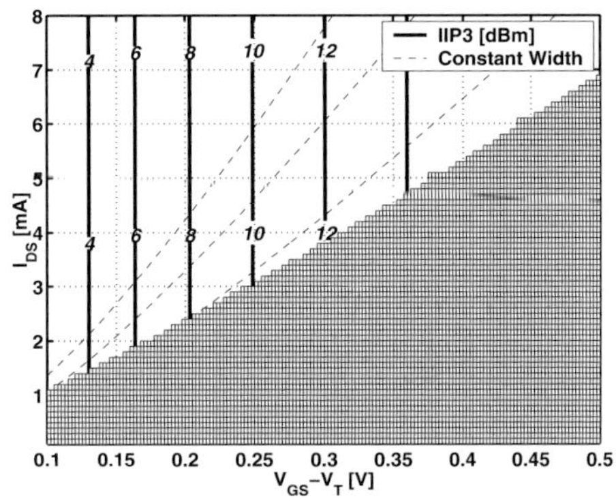

(d) IIP3 [dBm].

Figure 2.24: *Contour plots illustrating the behavior of the CG LNA.*

source. An extra coupling capacitor can be placed in series with this resistor in order not to disturb the DC bias. This resistance will lower the spread of the input impedance on process variations. However, it will also reduce the gain and increase the noise figure and should not be done unless it is really necessary.

Equation (2.118) is illustrated graphically in Fig. 2.24(a). The patterned region indicates the part of the design space where no input match can be obtained. The required load resistance increases towards the upper left corner of the graph since this region features a low $V_{GS} - V_T$, increasing the positive term in (2.118), and a high current, reducing the negative term. Moreover, if g_m is significantly larger than proscribed by (2.119) the second term in (2.118) can be neglected making R_L a pure function of $V_{GS} - V_T$. This is seen in the steep behavior of the 1500 Ω contour. It should also be mentioned that for the common source LNA discussed previously, R_L was assumed constant and was set to 500 Ω. If higher values for R_L are possible then the gain of the CS LNA can be increased[3]. If not, the upper left region in Fig. 2.24 should be patterned since

[3] Care has to be taken since high gain values may compromise the stability of the amplifier.

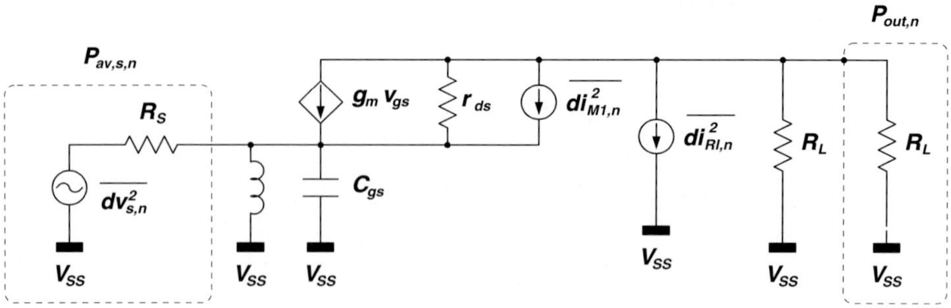

Figure 2.25: *Simplified small signal schematic of the CG LNA in Fig. 2.22 with noise sources.*

no input match can be obtained. $R_{L,max}$ is dependent on the technology. It depends on the size feature of the technology, but also on the amount of metal layers, their sheet resistance and the distance between them. It usually has an upper bound of 1 to 2 kilo Ohm. This will be discussed in more detail in Section 4.8.

2.6.2.2 Power Gain

Assuming the input of the LNA is matched, the output current, equal to the input current is given by

$$i_{out} = i_{in} = \sqrt{\frac{P_{av,s}}{R_S}} \tag{2.120}$$

A derivation similar to the one in Section 2.6.1.2 results in the following expression for the power gain:

$$G_T = \frac{R_L}{4R_S}. \tag{2.121}$$

Substituting (2.116) in (2.121) yields

$$G_T = \frac{1}{2}\left(ng_m - \frac{1}{R_S}\right) r_{ds}. \tag{2.122}$$

The result is plotted in Fig. 2.24(b). Comparing Fig. 2.24(a) and Fig. 2.24(b) clearly shows that the gain increase towards the upper left corner solely results from an increased load resistance. Also note that the power gain is much lower than for the CS LNA. Typical values are in the range of 5 to 10 dB compared to 25 dB for the CS LNA.

2.6.2.3 Noise Figure

Calculation of the noise factor of the CG LNA is based on the schematic in Fig. 2.25. Only the classical drain channel noise and the thermal noise of the load resistance are considered. The noise factor is approximated by

$$\begin{aligned} F &\approx 1 + (F_d - 1) + (F_L - 1) \\ &\approx 1 + \frac{\gamma g_m}{\alpha} \frac{4r_{ds}^2 R_S}{(2r_{ds} + R_L)^2} + \frac{4R_S}{R_L}, \end{aligned} \tag{2.123}$$

where $F_L - 1 = G_T^{-1}$. Substitution of (2.116) in (2.123) yields

$$F \approx 1 + \frac{\gamma}{ng_m R_S} + \frac{2R_S}{(ng_m R_S - 1)\, r_{ds}}. \tag{2.124}$$

It is seen that $(F_d - 1)$ can theoretically be made arbitrarily low by simply increasing the transconductance. This is a big advantage of the CG amplifier. The performance of the CG LNA with respect to noise figure and gain is lower than that of the CS LNA at low frequencies. However for the CS LNA the excess noise is for the most part proportional to the square of the operation frequency and the power gain is inversely proportional to the square of the frequency. Consequently, the performance rapidly declines with increasing frequency. And at higher frequency the CG LNA performs better then its CS counterpart [Gua02]. Since this comparison requires a more detailed performance model taking into account non-quasistatic effects and several other parasitics and non-idealities it will be discussed in more detail in Section 4.11.

The noise figure of the CG LNA is presented graphically in Fig. 2.24(c). Similarly to the gain, also the noise figure improves towards the upper left, i.e. for larger g_m. This is easily understood from equation (2.124) since $(F_d - 1)$ is inversely proportional to g_m. It is quite interesting to see that the noise figure contours almost coincide with the contours for R_L. Intuitively this may be understood as follows. Since the input is matched, half of the source noise current is delivered to the output due to the nature of the matched CG stage. However, the channel noise current is not necessarily found at the output. Part of the current will simply run through r_{ds}. Suppose R_L were infinite then none of the channel noise current would find its way to the output; it would all flow through r_{ds} instead. Hence, a higher R_L will lead to a lower $(F_d - 1)$. Also $F_L - 1 = G_T^{-1}$ is only dependent on R_L.

2.6.2.4 Linearity

Analogously to the discussion in Section 2.6.1.4, the IIP3 of the CG LNA expressed in [dBm] is approximated by

$$\text{IIP3} = 11.25 + 10\log\left(\frac{V_{GST}\,(2 + \Theta V_{GST})\,(1 + \Theta V_{GST})^2}{\Theta}\right). \tag{2.125}$$

Compared to (2.110) the last term $(-20\log(Q_{in}))$ is replaced with $+6$ dB since $v_{gs} = v_s/2$ instead of $v_{gs} = Q_{in} v_s$.

The result can be interpreted graphically by means of Fig. 2.24(d). Since there is no input quality factor, IIP3 is only function of $V_{GS} - V_T$ like for a simple MOS transistor. Indeed, the IIP3 contours in Fig. 2.24(d) are vertical lines. The absence of Q_{in} also allows much larger values for IIP3 compared to the CS LNA.

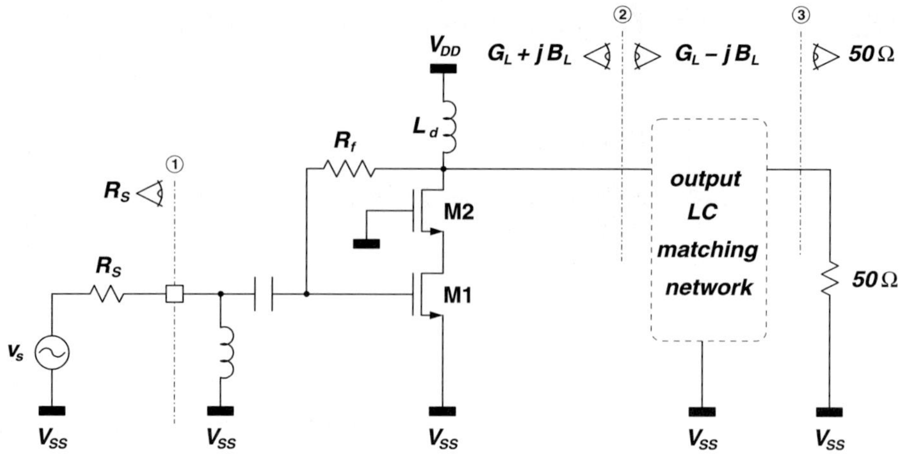

Figure 2.26: *Simplified schematic of the shunt feedback LNA.*

2.6.3 Shunt-Feedback Amplifier

A simplified schematic of a shunt-feedback LNA is shown in Fig. 2.26. The amplifier consists of a common source transistor with a cascode transistor on top to reduce the Miller effect. The resistor R_f is added to obtain a resistive part in the input impedance. The excess capacitance is tuned out with a parallel inductor, similar to the common-gate topology. If the feedforward through R_f can be neglected, then the input impedance is given by

$$Z_{in} = \frac{R_f}{1 + g_m \frac{R_{out}}{2}}, \tag{2.126}$$

where the output resistance R_{out} is found by

$$R_{out} = R_L \parallel R_f. \tag{2.127}$$

For a given g_m and R_L, R_f is chosen such that $Z_{in} = 50\ \Omega$. For $R_L = 1\ \text{k}\Omega$ and a frequency of 1.5 GHz, the contours of R_f are plotted in the design space of the amplifying transistor in Fig. 2.27(a). The patterned area marks the region where the power gain is negative.

The power gain of the shunt-feedback LNA is calculated by

$$G_T = \frac{g_m^2 R_S R_L \parallel R_f}{4}. \tag{2.128}$$

This is illustrated in Fig. 2.27(b). Indeed, the gain drops going to the lower right corner. In that region g_m is lowest and (2.126) implies that R_f needs to decrease in order to keep $Z_{in} = 50\Omega$. Thus the power gain drops even faster. A power gain of 15 to 20 dB is achievable depending on the power budget.

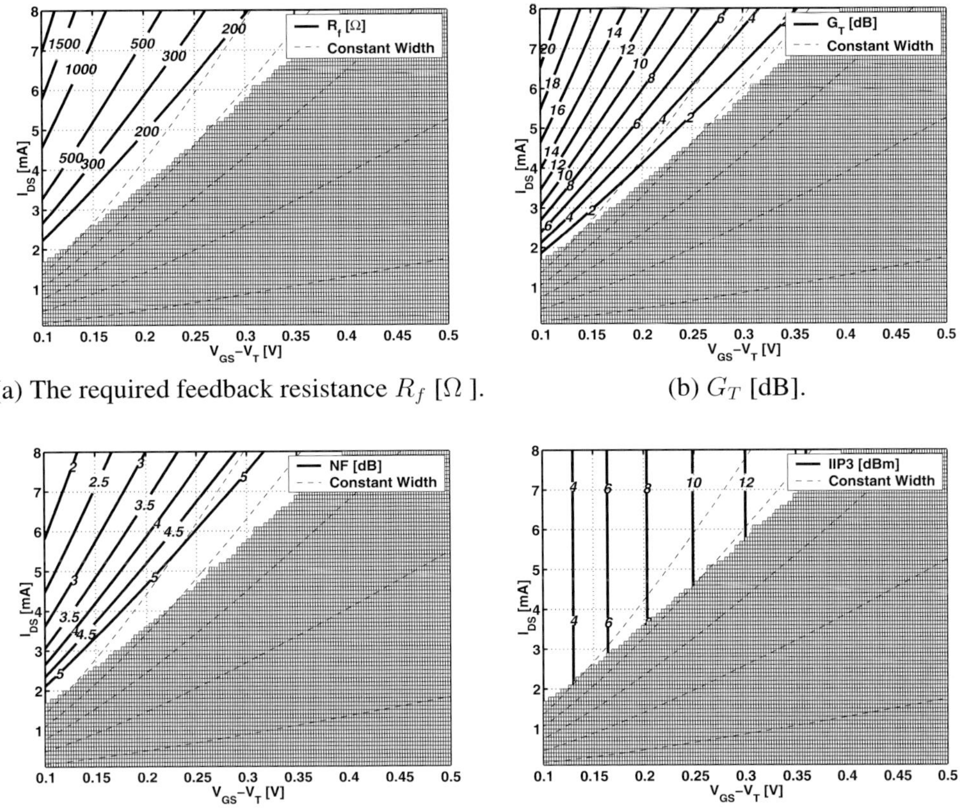

(a) The required feedback resistance R_f [Ω].

(b) G_T [dB].

(c) NF [dB].

(d) IIP3 [dBm].

Figure 2.27: *Contour plots illustrating the behavior of the shunt feedback LNA.*

The noise factor of the shunt-feedback LNA can be written as

$$F \approx 1 + \frac{R_S}{R_f} + \left(\frac{\gamma}{\alpha} g_m R_{out} + 1\right) G_T^{-1}. \tag{2.129}$$

The second term in this equation represents the contribution of the feedback resistor noise. It is usually the dominant contribution. Consequently, the noise figure can only be sufficiently low in the region where R_f is high according to (2.126). This is illustrated graphically in Fig. 2.27(c). Indeed, NF decreases towards the upper left. The noise performance is clearly inferior to both previous amplifiers.

The linearity of this type of amplifier is comparable to that of the common gate amplifier since their is no passive voltage amplification as with the tuned common-source amplifier. The

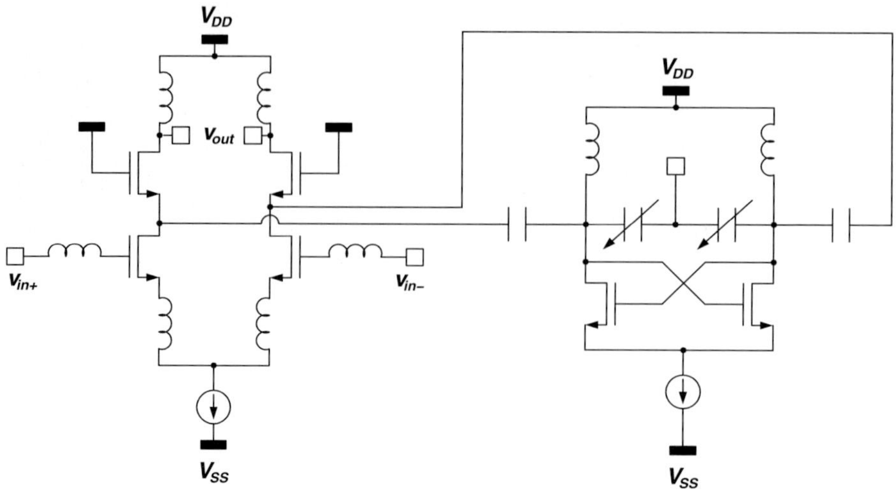

Figure 2.28: *Simplified schematic of an image reject LNA [Sam99].*

IIP3 is approximated by:

$$\mathrm{IIP3} = 11.25 + 10\log\left(\frac{V_{GST}\left(2+\Theta V_{GST}\right)\left(1+\Theta V_{GST}\right)^2}{\Theta}\right). \tag{2.130}$$

It is visualized in Fig. 2.27(d). This plot is identical to the common-gate amplifier. IIP3 is only function of $V_{GS} - V_T$.

One drawback of this circuit is the rather poor reverse isolation due to the direct connection of R_f to both input and output. It is in the order of 10 to 20 dB which is often too high for direct conversion receivers. Furthermore a small revers isolation will complicate the design for stability which becomes more problematic at high frequencies. Consequently this type of topology is only found at moderate frequencies.

An advantage of the shunt-feedback LNA is that, similar to the common-gate LNA, it can be used in baseband or wideband applications. The inductors used for tuning at the input and output are then left out. The output matching network is usually left out or replaced by an active buffer. Both amplifiers are for instance often used in optical receiver front-ends. These amplifiers are driven by a capacitive current source. They are intended to convert this current to a voltage and therefore in this context they are called transimpedance amplifiers.

In the remainder of this section, a few LNA designs or topologies will be discussed that don't completely fit within the foregoing classification.

2.6.4 Image Reject LNA's

Image reject LNA's have an extra functionality besides amplifying the wanted signal. They feature a sharp notch in their characteristic at the image frequency. The image frequency is

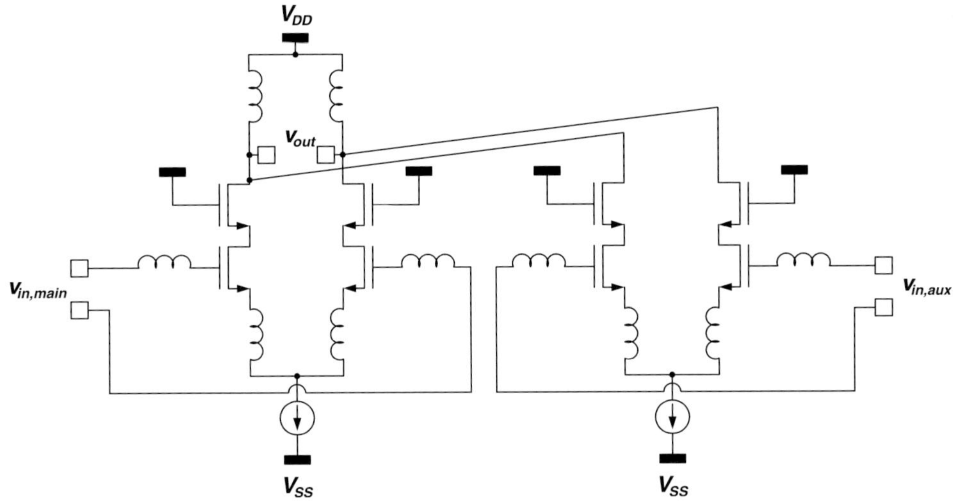

Figure 2.29: *LNA with feedforward cancellation of 3rd order distortion [Din01].*

located symmetrically to the wanted signal at the other side of the LO frequency. The notch is aimed at suppressing this image frequency.

One of the most interesting designs of an image reject LNA (and complete front-end) was done in [Sam99]. The basic schematic of the LNA + filter is shown in Fig. 2.28. The circuit was designed for a 5 GHz wireless LAN receiver and it was integrated within a complete RF front-end: LNA + image reject filter, PLL and mixer. The LNA is a differential, common-source amplifier with inductive degeneration. A series resonant tank is added at the cascode node, which shunts the signal to ground at the resonance frequency (the frequency of the image signal). The resonant tank that controls the notch is identical to the VCO in the PLL except for the larger supply current to sustain oscillation in the VCO. Further more, the center frequency of the notch is tuned by the same voltage that controls the VCO in the PLL. Consequently, locking the frequency of the VCO to the image frequency guarantees the suppression of the image frequency in the LNA.

2.6.5 Highly Linear Feedforward LNA

A highly linear LNA was presented at ISSCC in 2001 [Din01]. It achieves an IIP3 of 18 dBm through feedforward cancellation of the 3rd order terms. A schematic of the amplifier is shown in Fig. 2.29. It consists of a differential common-source amplifier with inductive source degeneration. A scaled copy of the amplifier is added in a feedforward configuration. This auxiliary amplifier is steered with a fraction $\beta > 1$ of the input signal. The output currents of both amplifiers are subtracted. The operation is as follows. Considering only 3rd order distortion, the output signal of the main amplifier can be written as

$$y_{main}(x) = a_1 x \left(1 + \frac{a_3}{a_1} x^2\right), \tag{2.131}$$

where

$$a_3 = \frac{4}{3} \frac{a_1}{\mathrm{IIV3}_{main}^2}. \tag{2.132}$$

For the auxiliary amplifier the output is given by

$$y_{aux}(x) = a_{1,aux} \beta x \left(1 + \frac{a_{3,aux}}{a_{1,aux}} \beta^2 x^2\right), \tag{2.133}$$

where

$$\frac{a_{3,aux}}{a_{1,aux}} = \frac{a_3}{a_1}, \tag{2.134}$$

since it is a scaled copy. Now if

$$a_{1,aux} = \frac{a_1}{\beta^3}, \tag{2.135}$$

then

$$y(x) = y_{main}(x) - y_{aux}(x) = a_1 \left(1 - \frac{1}{\beta^2}\right) x, \tag{2.136}$$

is a perfectly scaled version of x. Practically the linearity is limited by the mismatch between both amplifiers, higher order terms and non-linearity of the load impedance. A linearity improvement of 13 dB was achieved in [Din01]. However the comparison was done with an amplifier without feedforward but only using half the power. For the same power the improvement would be a few dB less.

2.6.6 The Noise-Cancelling Wide-band LNA

A very interesting LNA was introduced in [Bru02] at ISSCC 2002. The circuit is based on the shunt-feedback amplifier discussed in Section 2.6.3. The schematic is shown in Fig. 2.30(a). An extra inverting amplifier with voltage gain $-A_v$ is connected to the input. The output is added to the output of the shunt-feedback amplifier. The basic idea is that the input signal is 180 degrees out of phase with respect to the output signal while the input noise voltage is in phase with the output noise voltage. Consequently if the total input signal is fed forward through a (noiseless) inverting amplifier (with correct gain) and added to the output, then the noise of transistor M1 is completely cancelled while the signal is amplified.

2.6.7 Current Reuse LNA with Interstage Resonance

The current reuse topology implies that the LNA consists of two amplifier stages and only one current branch. A 5.2 GHz LNA employing this technique was published at ESSCIRC 2002 [Cha02, Cha03]. The basic circuit schematic is shown in Fig. 2.30(b). Operation is as follows.

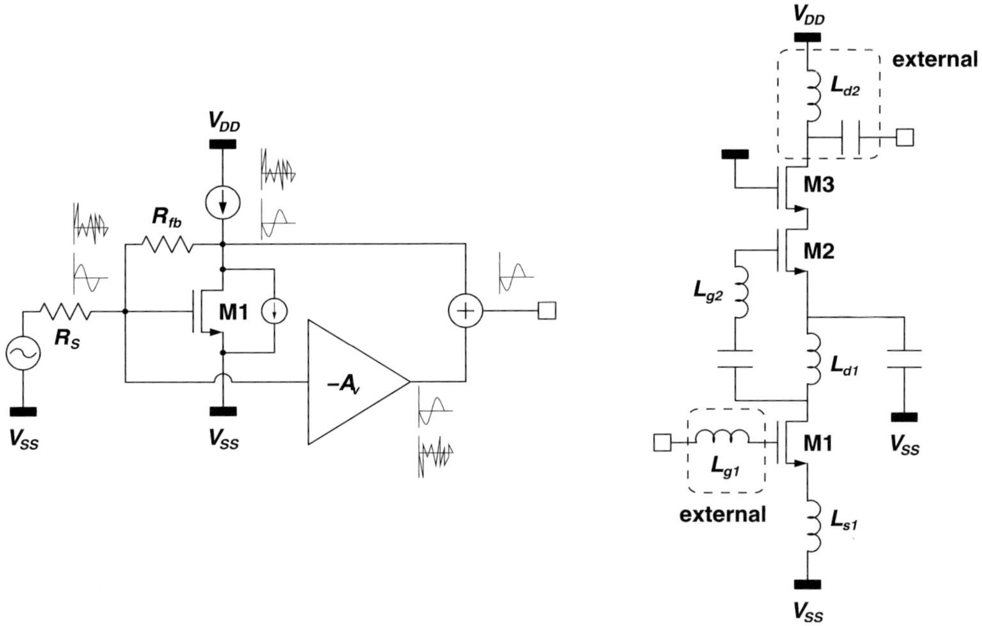

(a) [Bru02] (b) [Cha03]

Figure 2.30: *(a) Schematic of the noise-cancelling wide-band LNA [Bru02] and (b) schematic of the current reuse LNA with interstage resonance [Cha03] (b).*

Transistor M1 is a common source amplifier with inductive degeneration and a tuned load impedance. The voltage at the drain of M1 is passively amplified by the series resonance network consisting of L_{g2} and C_{gs2}. This v_{gs2} is then converted into a current by M2 and via cascode M3 the current is dumped in a second tuned load impedance.

2.6.8 Transformer Feedback LNA

Another design which is more than interesting is a transformer based common source LNA [Cas03]. A schematic of the circuit is shown in Fig. 2.31. The LNA was designed for application in a 5 GHz wireless LAN receiver. The circuit is fully differential and uses magnetic coupling between the input and output to reduce the Miller effect. This is achieved by interwinding the source and drain inductors in both branches. The Circuit is matched at the input and achieves a power gain of 14.2 dB. The noise figure of 0.9 dB is the lowest value published for a CMOS amplifier at this frequency.

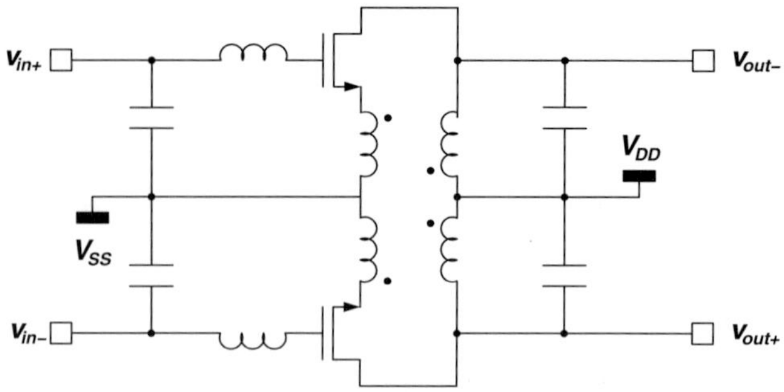

Figure 2.31: *Simplified schematic of the transformer feedback LNA [Cas03].*

2.7 Conclusion

In this chapter different concepts and topics, relevant to RF CMOS design have been introduced. A manageable MOS transistor model has been introduced that is able to reproduce the typical behavior of the device during design optimization. This model is closely related to Hspice MOS model level 3. The MOS model used in simulations has been extended with 10 extra resistors and capacitors to obtain more realistic simulation results with respect to noise, gain and linearity. Also the relevant non-quasistatic effect at the gate of the MOS device is modelled by this. All extra components are calculated taking into account the physical layout of the devices.

After a study of the different CMOS noise sources, the functionality of the LNA within the receiver chain has been studied. Based on the twoport cascading theory the conclusion was drawn that the LNA sets a minimum on the attainable noise figure of the complete receiver. Similarly, it places an upper bound on the attainable IIP3. The importance of the power gain and voltage gain has also been demonstrated. The value is determined by the dynamic range of the subsequent mixer stage. Also other LNA requirements have been investigated and placed in perspective: matching, reverse isolation, and stability.

The last section was devoted to the introduction of the most common LNA topologies. Even though no parasitic effects have yet been considered already a few interesting conclusions could be drawn. The common-source LNA with inductive degeneration seems to yield the best performance with respect to noise figure and gain. Unfortunately the noise figure increases rapidly with rising frequency. The common gate topology does not suffer from this effect. Moreover, the noise figure of the CG LNA can be improved continuously by increasing the current consumption. It may therefore become interesting to shift to this topology when the frequency and power budget are large. Both the CG amplifier and the shunt-feedback amplifier have the additional ability to handle baseband signals. This chapter concluded with a brief discussion of a few specific and interesting designs published recently.

Chapter 3

ESD Protection in CMOS

3.1 Introduction

Electrostatic Discharge or ESD is the discharge of electrostatic charge or energy. The name is rather peculiar because it presents a "contradictio in terminis". Electrostatics implies the distributed presence of electrical charge giving rise to a distributed but constant electric field. Both charge and electric field are distributed in space but constant in time. A discharge implies that the charge is flowing from one place to another. Hence, the charge distribution changes in time. This is in contradiction with the very definition of electrostatics. A discharge can never be static, but is always dynamic. A more appropriate term would be the "dynamic discharge of electrostatic charge". But what's in a name?!

The world of ESD is very familiar to many. Who hasn't taken of his sweater and seen or felt the typical sparks? The voltages associated with these sparks can be extremely large and easily reach tens of thousands of Volts. Even though this seems exceedingly dangerous, the energy associated with the sparks is so small that they are seldom more than an annoyance to humans. Unfortunately they can be devastating for an IC.

Several types of damage can be incurred by an electrostatic discharge into the pins of an IC. The most common failure originates from damage to the gate oxide. This may happen when the gate of a MOS transistor is directly connected to a pin of the IC which is quite common. The discharge current increases the voltage over the gate oxide and destroys it when the breakdown voltage of the oxide is reached. This breakdown voltage scales linearly with the oxide thickness which in turn scales linearly with the feature size of the technology. Hence the devices become more and more sensitive to ESD.

Not only the gate oxide but also the inherent pn-junctions are not impervious to the ESD hazard. An increasing reverse voltage will increase the thickness of the junction but not enough to keep the electric field from rising rapidly within the depletion region. Under the influence of this large field, a lot of electron-hole pairs are generated which reach saturation velocity quickly. If the electric field is sufficiently high, collisions will activate even more electron-hole pairs. This vicious circle is suitably named avalanche breakdown. It will cause the temperature of the silicon to rise. Eventually the silicon will start to melt. This is called second breakdown. The incurred

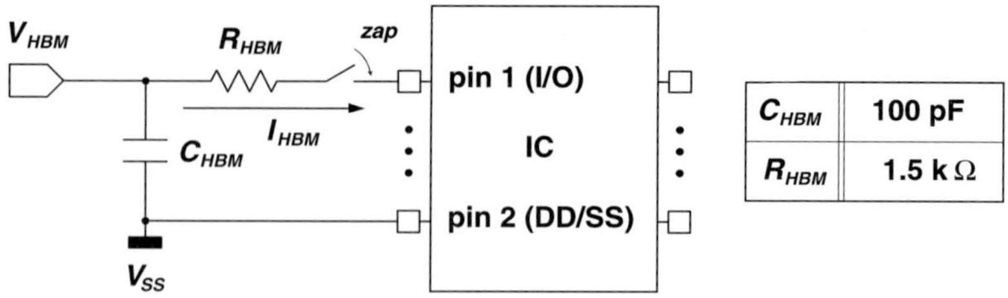

Figure 3.1: *Setup for Human Body Model testing.*

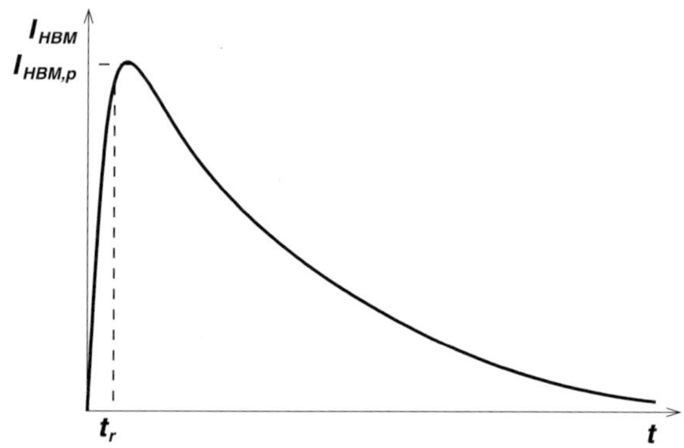

Figure 3.2: *Typical HBM discharge current.*

damage is irreversible. The device is destroyed.

Several ESD protection elements have been described in open literature: diodes, silicon controlled rectifiers (SCR), grounded gate NMOS (GGNMOS), gate coupled NMOS (GCNMOS), etc.. They are discussed separately in Section 3.3. Different ESD-protection tests have been devised and several standards have been developed that are used in industry. They are explained first in Section 3.2

3.2 ESD Tests and Standards

3.2.1 Human Body Model

The Human Body Model is intended to represent a human discharging into an IC. It is defined by the ESD Association Standard 5.1 [inb93]. A very basic schematic of the setup for HBM testing is shown in Fig. 3.1. The tester is connected to pin1 which will be zapped with a positive pulse versus pin2. Pin2 is connected to ground. C_{HBM} is the capacitor which is initially charged by a high voltage source to V_{HBM}. It is then discharged through R_{HBM}, a 1.5 kΩ resistor, into the IC.

HBM Voltage level	Peak current	Rise time	Decay time	Satisfied class
250 V	0.15-0.19 A	2-10 nS	130-170 ns	Class Ia
500 V	0.30-0.37A	2-10 nS	130-170 ns	Class Ib
1000 V	0.60-0.74 A	2-10 nS	130-170 ns	Class Ic
2000 V	1.20-1.48 A	2-10 nS	130-170 ns	Class II
4000 V	2.40-2.96 A	2-10 nS	130-170 ns	Class IIIa
8000 V	4.80-5.92 A	2-10 nS	130-170 ns	Class IIIb

Table 3.1: *Overview of the HBM test indicating the different classes and the main pulse features.*

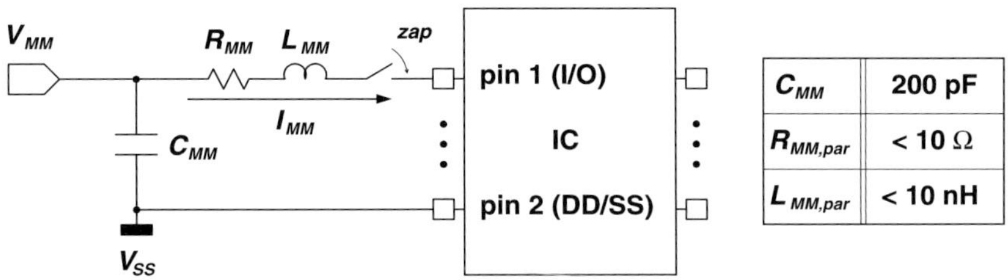

Figure 3.3: *Setup for Machine Model testing.*

Fig. 3.2 plots the typical behavior of the HBM discharge current. The peak current is 0.67 A/kV which is simply $1/R_{HBM}$. The rise time of the HBM pulse is in the order of 5 nS. The main frequency content is situated in the range of a few MHz which is relatively low compared to the Machine Model and especially the Charged Device Model discussed in Section 3.2.2 and Section 3.2.3. Since the HBM test has a very high discharge resistance, it is very insensitive to excess input inductance which makes it easy to implement.

Several industrial standards exist that are based on the HBM test. They are expressed as the HBM-voltage the IC can withstand. The most commonly used standard is 2 kV which is indicated as Class II in Table 3.1. This means the peak current the ESD-protection should be able to sink is 1.33 A while limiting all internal voltages to a safe level, sufficiently below breakdown. The ESD Association standards also specify that each I/O pin should be zapped versus each power supply pin. Shorting all power supply pins together is allowed but not always convenient, especially with manual testers. Three repeated zaps in sequence with a time interval of 1 second minimum are required for the circuit to pass the test.

3.2.2 Machine Model

The Machine Model represents charging due to machine handling. It is primarily used in Japan and in the automotive industry and is covered under specification [inb94]. Fig. 3.3 illustrates the basic MM setup. It is quite similar to the HBM setup. In the MM test however, current peaks will be larger and frequency components will be higher than with the HBM test. This is due to

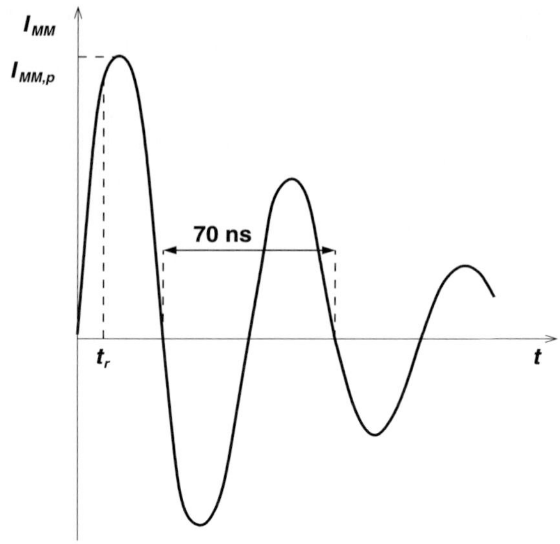

Figure 3.4: *Typical MM discharge current.*

MM Voltage level	Peak current	Resonance frequency	Satisfied class
100 V	1.5-2 A	11-16 MHz	Class A
200 V	2.8-3.8 A	11-16 MHz	Class B
400 V	5.8-8 A	11-16 MHz	Class C

Table 3.2: *Overview of the MM test indicating the different classes and the main pulse features.*

the much smaller discharge resistance. Moreover, the RLC nature of the discharge will render the currents a rather oscillatory behavior compared to the RC nature of the HBM pulses. The main pulse characteristics can be found in Table 3.2. The resonance frequency is typically about 15 MHz and the peak current is is in the order of 17 A/kV, over 20 times larger than for the HBM test. Table 3.2 also shows the different standard classes. Most common standards for IC's are Classes A and B.

3.2.3 Charged Device Model

The Charged Device Model (CDM) is quite different from the previous two models. Where as both previous models describe what happens when one particular pin is zapped versus another pin, the Charged Device Model represents the self-charging and self-discharging of the complete die. Specifications of the standard can be found in [inb99]. A schematic for realization in the laboratory is shown in Fig. 3.5. A large charging plate is connected through a switch to a high-

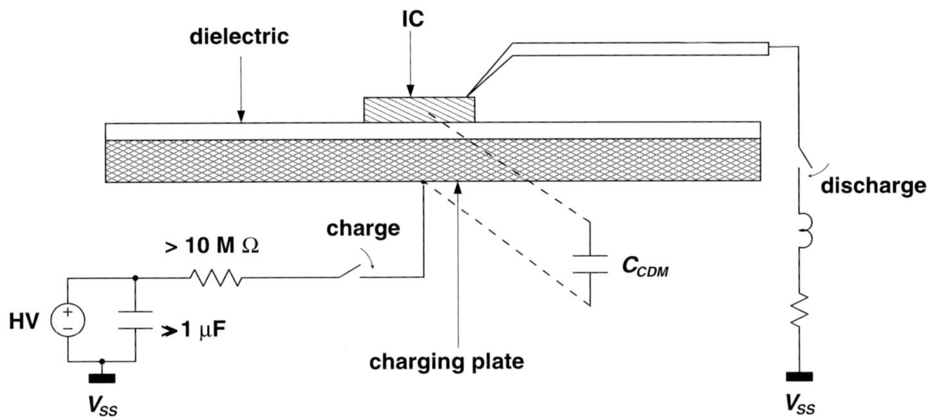

Figure 3.5: *Setup for Charged Device Model testing.*

voltage source. The plate is galvanically shielded from the chip by a thin dielectric. Hence, a capacitor is formed between the plate and the IC. Two different plate sizes have been standardized giving a coupling capacitance of either 6.8 pF or 55 pF. One of the pins of the IC is connected through a switch and some parasitic inductance and resistance to ground.

The test comprises two phases. The first phase is the field-induced charging. Hereby the charging plate is charged to the wanted CDM voltage. Since the IC is electrically floating and capacitively coupled with the charging electrode, it will be charged to the same CDM voltage. In the second phase, one of the pins of the IC is shorted to ground using a 1 Ω probe with some stray inductance. This causes a very swift discharge of the IC. A typical CDM discharge current is plotted in Fig. 3.6.

Due to the very low discharge resistance, the CDM pulse has the fastest transients and has the maximum peak current of the discussed models. It is therefore usually the hardest to protect against. Also the effect of parasitics, both inductive, capacitive and resistive is most pronounced in the CDM test. An overview of the most important CDM pulse features for both disk sizes can be found in Table 3.3. Table 3.4 denominates the different classes in the CDM standard. For an IC, the most common standard is class II.

3.2.4 Transmission Line Pulsing

The previous ESD tests tried to emulate the ESD stress as it would occur in the chip's "natural habitat" (human contact for HBM, machine contact for MM). Another strategy to do ESD stressing is based on the use of a very simple stress tool. Afterwards, the resulting data can be extrapolated to ESD conditions. In the test discussed here, a transmission line will be used as a current source (actually a charge source). This test will yield a very repeatable set of measurements which can be used to characterize a specific device or IC. It is a lot less influenced by

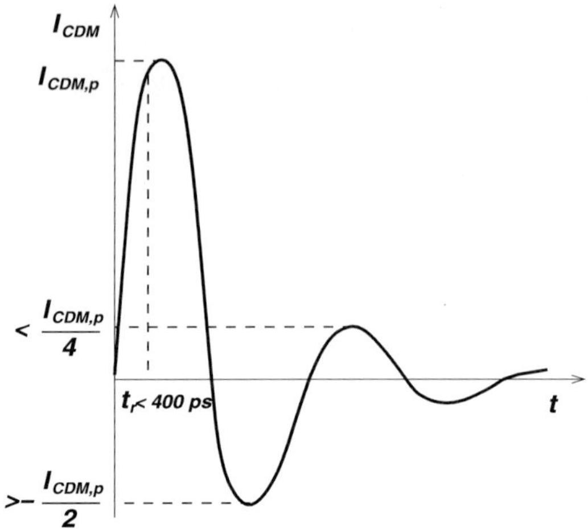

Figure 3.6: *Typical CDM discharge current.*

CDM Voltage level	Disk size	Capacitance	Peak current	Rise time
200 V	large	55 pF	4.5 A	-
500 V	large	55 pF	11.5 A	-
500 V	small	6.8 pF	5.75 A	<400 ps
1000 V	small	6.8 pF	11.5 A	<400 ps

Table 3.3: *Overview of the CDM test indicating the main pulse features.*

CDM Voltage level	Satisfied class
<200 V	Class I
>200 V	Class II
>500 V	Class III
>1000 V	Class IV

Table 3.4: *Overview of the CDM test indicating the different classes.*

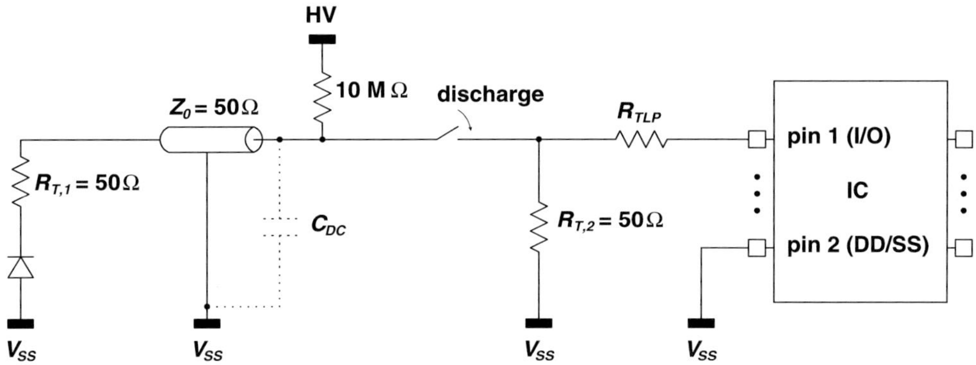

Figure 3.7: *Setup for TLP testing.*

parasitic effects caused by the exact length or placement of the wires etc..

The setup for the Transmission Line Pulse (TLP) test is illustrated in Fig. 3.7. A transmission line with characteristic impedance $Z_0 = 50\ \Omega$ is charged to a high voltage through a large resistance (10 MΩ). The amount of charge in the transmission line is related to its DC capacitance according to

$$Q_{TLP} = V_{TLP} \times C_{DC}. \tag{3.1}$$

C_{DC} itself is proportional to the length of the transmission line. The resistor R_{TLP} is intended to increase the output impedance of the tester in order to make it appear as a current source. Its value is usually 500 Ω or more. Resistors R_{T1} and R_{T2} are used to terminate the transmission line at both ports to avoid internal reflections. The reverse diode prevents leakage of the charge to ground through R_{T1} during charging. Its junction capacitance should be large enough such that its impedance is much lower than 50 Ω at the frequency where transmission line behavior starts [1].

A typical TLP test is based on the following scenario. First the leakage current between the two pins which are going to be stressed, is measured. After that the transmission line is charged to a certain voltage (the initial voltage should be relatively low such that the IC is sure to survive the test). The transmission line is subsequently discharged into the IC. At that point, the current into the IC and the voltage over the pins can be visualized and recorded with an oscilloscope. After discharge, the leakage current is measured again. These steps are repeated with a gradually increasing TLP voltage until there is a significant change in the intermediate leakage current. The obtained maximum TLP voltage, input current and pin voltages can be used to extrapolate the behavior of the IC under real ESD-stress.

[1] Transmission line behavior starts when the length of the transmission line is about one tenth of the wavelength in the transmission line.

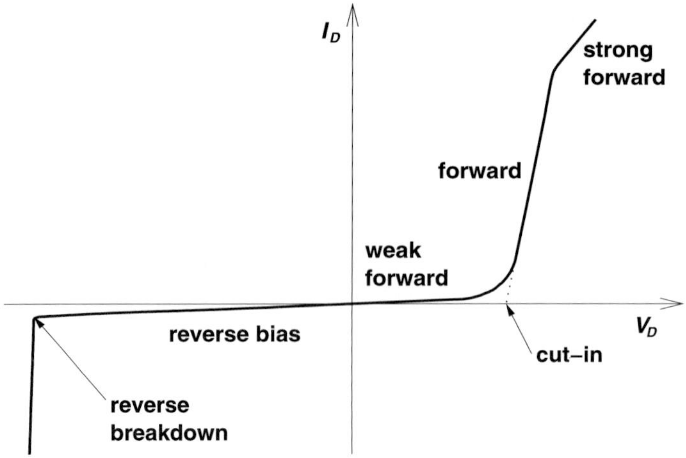

Figure 3.8: *Diode IV characteristic.*

3.3 ESD-Protection in CMOS

3.3.1 ESD-Protection Devices

3.3.1.1 Diode

The diode is one of the most frequently used ESD-protection devices. The main reason is that they are very efficient and robust. Furthermore, their characteristics are fairly simple to model and simulate, allowing a reliable sizing of these devices. This is especially important if the size of the ESD device directly affects the performance of the core circuit, as is the case for RF IC's. The diode has four basic regions of operation: reverse bias, weak forward bias, forward bias and strong forward bias (see Fig. 3.8). The forward bias of an ideal diode is described by the equation:

$$I_D = I_{D0}(e^{V_D/V_t} - 1), \tag{3.2}$$

where I_{D0} is the saturation current, V_D is the forward voltage over the diode, and $V_t = kT/q$ (26mV at room temperature). Hence, the forward voltage of a diode is given by

$$V_D = V_t \times \ln\big(\frac{I_D}{I_{D0}} + 1\big) + IR_D, \tag{3.3}$$

where R_D is the series resistance in the diode. It is clear that the latter term may not be neglected in an ESD context since currents may go up to a few Amps. The +1 term may almost always be neglected, even for relatively weak forward bias.

For an ideal diode in the forward region, the voltage will decrease as the temperature is

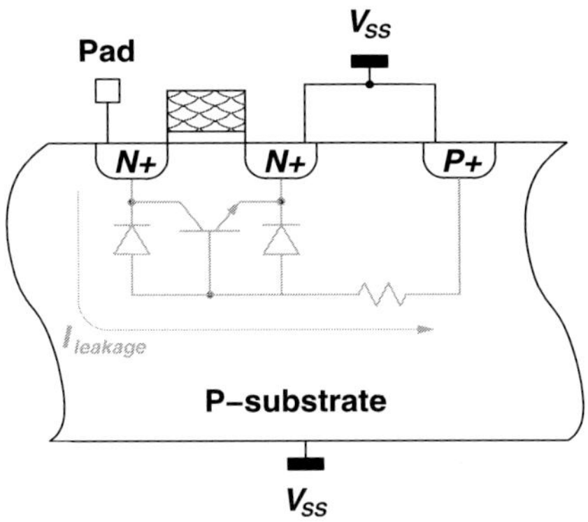

Figure 3.9: *Cross section of a grounded-gate NMOS, also indicating the parasitic devices.*

increased from T_0 to T, described by the following linear relation:

$$V_D(T) = nE_{g0} + \frac{T}{T_0}\left(V_D(T_0) - nE_{g0}\right), \tag{3.4}$$

where $E_{g0} = 1.206\ V$ is the silicon bandgap and n is the diode ideality factor. These equations are valid only for ideal diode and get more complex for real CMOS diodes where they depend strongly on the specific type and process.

3.3.1.2 Grounded-Gate NMOS

A Grounded-Gate NMOS device (GGNMOS) is formed by shorting the gate of an NMOS transistor to ground. A cross section of the device is shown in Fig. 3.9. Since it consists only of a standard NMOS transisor, the device is fully compatible with any CMOS technology. No thick field oxide is required. The gate of the device is shorted in order to ensure that the NMOS is off at all times. The operation in case of an ESD event is based on the snapback mechanism. This means that at a certain voltage and current level, the parasitic bipolar transistor is turned on and starts to sink the ESD-current. This parasitic npn transistor is formed by the n+ drain contact, the p- substrate and the n+ source contact.

The snapback action is clarified in Fig. 3.10. Suppose a positive ESD-pulse arrives at the input pad (note that generally there is not a whole lot positive about an ESD-pulse, except its polarity, as is implied here). Since the pad is directly connected to the drain of a GGNMOS, the drain voltage increases and so does the reverse leakage current of the n+drain p-sub junction. This causes the p-substrate voltage to go up until the p-sub n+source junction becomes forward biased. At this point, the parasitic npn transistor is activated and rapidly starts to take a lot of the ESD-current. As a consequence the collector or drain voltage (and therefore the pad voltage) is drastically lowered. If the total ESD-current increases further, the holding voltage increases also depending on the on-resistance of the device. If the current increases beyond the I_{t2} of the device, it goes into second breakdown and is destroyed.

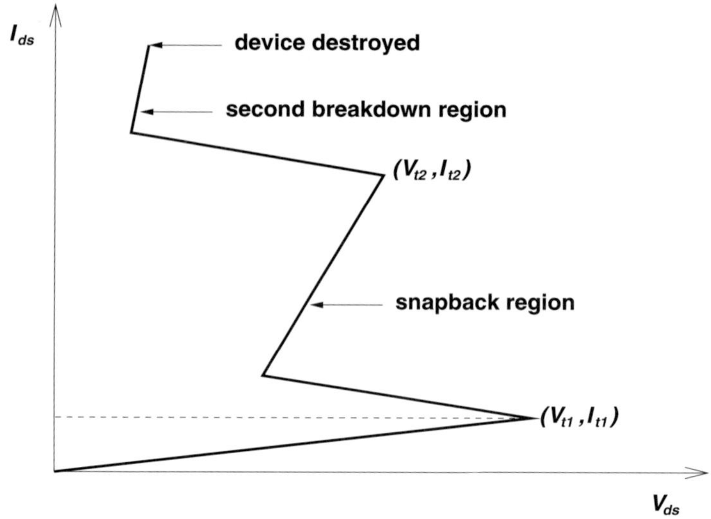

Figure 3.10: *Snapback action in a grounded-gate NMOS.*

During snapback operation, the current flows uniformly over all fingers and hence the current scales linearly with the width of the device. This scaling is possible since the on-resistance during snapback has a positive temperature coefficient. This means that if the current in a certain path increases, the temperature will rise and the resistance will go up. This will cause the current to favor other paths hence uniformly distributing the current. However, close to the onset of second breakdown, the on-resistance features a negative temperature coefficient causing the current to favor the already overloaded current path.

In order to avoid the latter scenario, the snapback voltage needs to be low compared to the second breakdown voltage. This will ensure that all fingers go into snapback before second breakdown is initiated in any one of them. Three strategies are very useful:

- Increase the V_{gs} during ESD, which has been shown to decrease the snapback voltage [Pol92].
- Reduce the channel length to decrease the snapback voltage since the β of the npn increases due to its smaller base.
- Add ballast resistance.

The first strategy is discussed briefly in Section 3.3.1.3. The second one is intuitively clear but the latter may require some clarification.

The influence of adding a drain resistor ballast, or simply ballasting, may be understood by examining a two finger GGNMOS as depicted in Fig. 3.11. Fig. 3.11(a) shows the two finger device without ballasting. Again , say a positive pulse arrives at the pad. The voltage at the pad, $V_{pad} = V_{d1} = V_{d2}$ increases. Now suppose finger 1 goes into snapback first. The voltage over the

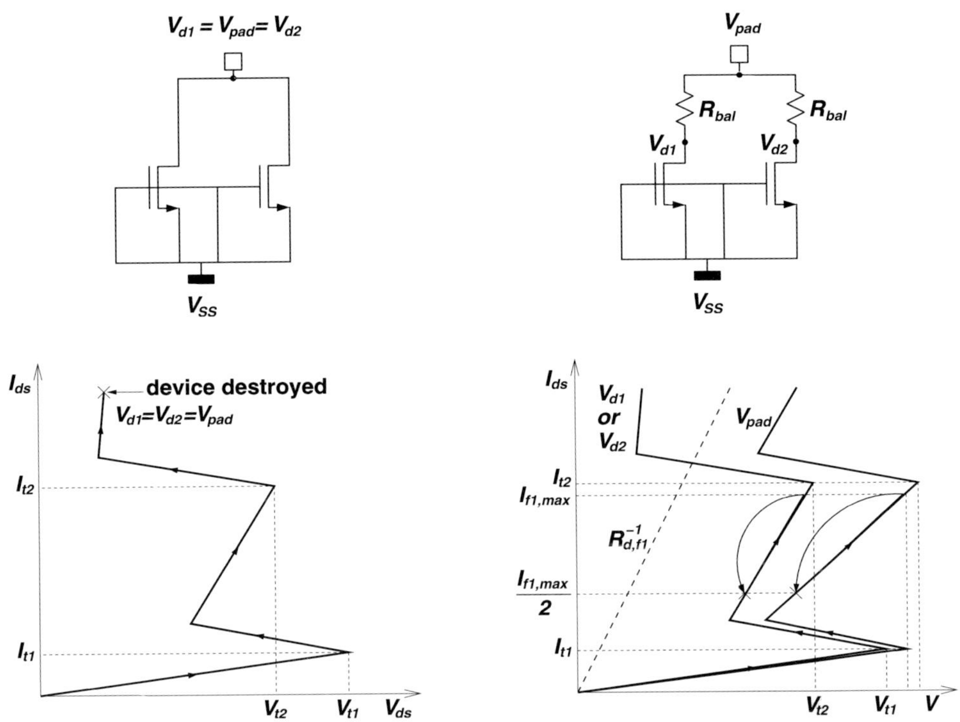

(a) No ballasting. (b) With ballasting.

Figure 3.11: *IV behavior of a two-finger GGNMOS under ESD-stress.*

device will drop because finger 1 will take a lot of current. Since V_{d2} is reduced, finger 2 does not go into snapback. If the current increases further, finger 1 will reach the second breakdown voltage before finger 2 has a chance to snapback since the second breakdown voltage, $V_{t2}$is lower than the snapback voltage V_{t1}.

Now what happens if a drain ballast resistor is added in each finger. This is illustrated in Fig. 3.11(b). Again suppose the same pulse arrives at the pad and finger 1 goes into snapback first. This happens when $V_{d1} = V_{t1}$ and $V_{pad} = V_{t1} + R_{bal} I_{t1}$. Since V_{d2} is again reduced, finger 2 does not go into snapback. If the current increases further, V_{d1} will increase but this time V_{pad} will increase faster. When $V_{pad} = V_{t1} + R_{bal} I_{t1}$, the second finger will snapback. At that point $V_{d1} < V_{t2}$ and no second breakdown can occur. As the second finger has now snapped back, the current is evenly distributed over both fingers. Specifically for finger 1 this means that the current that was flowing just before the second finger snapped back, $I_{d1,max}$ is halved, the other half is flowing in finger 2. Understanding the above allows the calculation of the ballast resistance per finger. It can be obtained from

$$V_{t1} + R_{bal} I_{t1} < V_{t2} + R_{bal} I_{t2}, \tag{3.5}$$

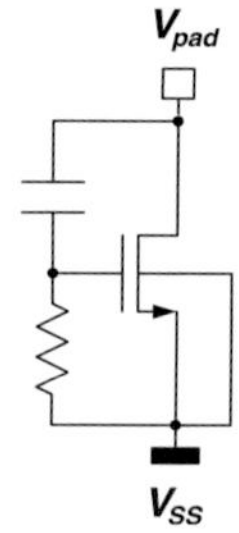

Figure 3.12: *Basic schematic of a gate-coupled NMOS.*

which states that the 'new' snapback voltage should be sufficiently lower than the 'new' second breakdown voltage [2]. This yields

$$R_{bal} > \frac{V_{t1} - V_{t2}}{I_{t2} - I_{t1}}. \tag{3.6}$$

Of course, increasing R_{bal} will result in a larger holding voltage which may harm the circuit, the GGNMOS needs to protect. For instance for a device with 10 fingers which should be able to take a current of 1 A. A resistance of 10Ω per finger will increase the holding voltage with 1 V.

3.3.1.3 Gate-Coupled NMOS

One of the strategies to lower the snapback voltage of an NMOS transistor, mentioned in Section 3.3.1.2 is to increase the V_{gs} during ESD. However the turn-on must be weak, or else the second breakdown current, will be degraded. This weak turn-on is realized by a capacitor which is used to couple a fraction of the ESD-charge to the gate of the protecting NMOS as shown in Fig. 3.12. The operation is based on a high-pass filter implemented by capacitor C_1 and resistor R_1 [Duv95]. As such, enough charge is coupled at the initial stage of the ESD-pulse such that the NMOS is weakly turned on.

3.3.1.4 Silicon-Controlled Rectifier

The Silicon-Controlled Rectifier (SCR) is also known as a thyristor. Even in standard CMOS it is possible to use this structure for ESD-protection. Its operation is explained by means of Fig. 3.13.

Assume again that a positive pulse arrives at the pad, the anode of the thyristor. As the anode voltage increases, the anode current increases slowly due to the leakage current $I_{leak,1}$ through the reverse-biased n-well/p-sub diode D_1. Then the anode current starts to increase noticeably when the anode voltage is increased above roughly 7 V, as a result of punchthrough between the n-well and the n+ cathode. The injected electrons from the cathode by punchthrough cause holes to be generated by impact ionization at the reverse-biased n-well/p-sub junction, which flow to

[2] Actually the snapback voltage should be lower than the voltage at which the temperature coefficient becomes negative which is somewhat smaller than V_{t2}.

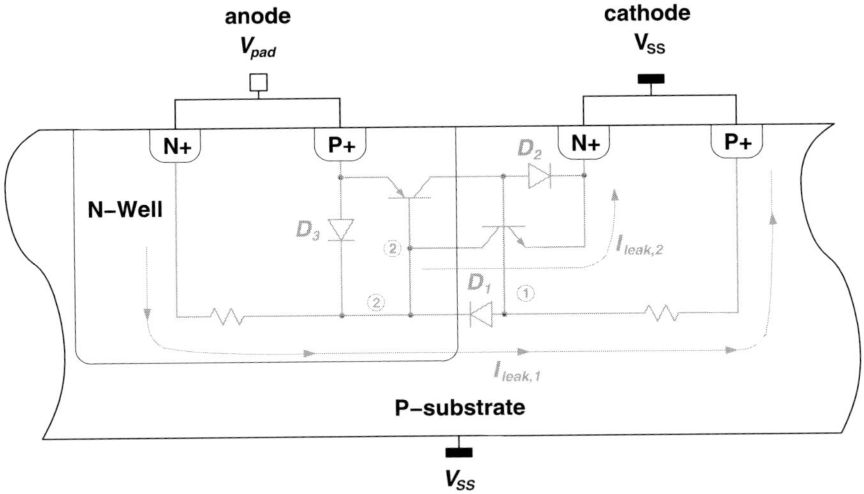

Figure 3.13: *Cross section of a Silicon-Controlled Rectifier, also indicating the parasitic devices.*

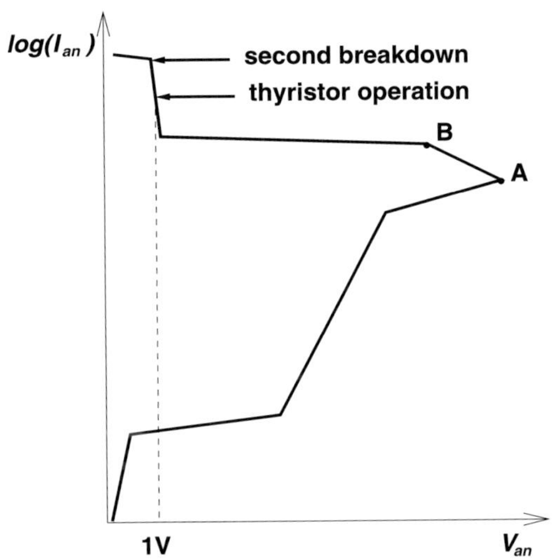

Figure 3.14: *IV-characteristic of the SCR.*

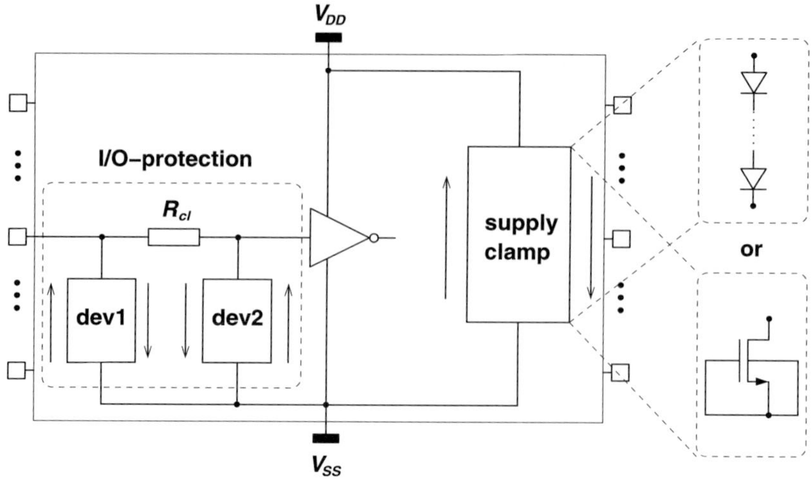

Figure 3.15: *Principal schematic of an IC indicating the I/O protection and the supply clamp.*

the p-substrate increasing the body potential at node ①. As the anode voltage increases, with sufficient hole current flowing, the body potential near the cathode junction gets high enough to forward-bias the p-sub/n+ cathode diode D_2 triggering the lateral npn (n+ cathode/p-sub/n-well) bipolar transistor. The n+ cathode, p- substrate, and n-well act as the emitter, base, and collector, respectively. At this situation, a snapback is monitored as indicated by point A in Fig. 3.14.

The bipolar current from the n+ anode flows through the n-well, which decreases the potential of the region under the p+ anode by ohmic drop. When the bipolar current is large enough, the p+ anode/n-well junction is forward biased to trigger the pnpn (p+ anode / n-well / p-sub / n+ cathode) thyristor, which causes another decrease in the anode voltage, as indicated by point B in Fig. 3.14. The resulting holding voltage drops to about 1 Volt, which is much smaller compared to that of the GGNMOS, treated in Section 3.3.1.2.

3.3.2 ESD-Protection Topologies

Where Section 3.3.1 discussed some of the more frequent devices used for ESD-protection, this section will treat the topologies in which the devices can be fitted. The first subsection will handle the topologies for I/O pins. The second subsection deals with topologies for power supply clamps.

3.3.2.1 I/O Pins

A very general schematic of an IC is shown in Fig. 3.15. Three devices can be recognized in the I/O protection circuit. Dev1 is the primary protection device. It should be a very large structure since its job is to take most of the ESD-current. The holding voltage of this device is not so important since the voltage over it is decoupled from the voltage at the gate of the inverter by

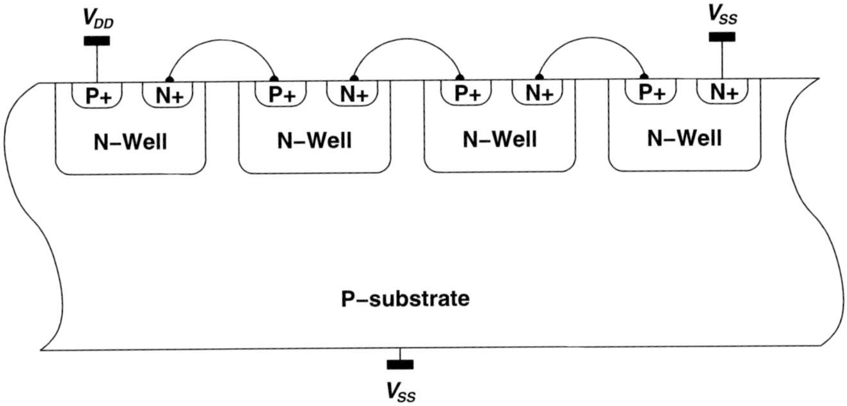

Figure 3.16: *Cascaded diodes for a supply clamp.*

means of current limiting resistor R_{cl}. The primary goal of this resistor is to limit the amount of ESD-current that is going towards the I/O circuit block (the inverter in Fig. 3.15). The value of the resistance can range from ten to a few hundred Ohm. It should clearly be taken into account in the functional design of the IC. For digital circuits, this resistance could drastically slow down the charging or discharging of the input node. For analog and RF circuits the influence of the noise should be considered as well. The third device in the I/O-protection is denoted dev2 in Fig. 3.15. Its purpose is to make sure that the voltage at the input of the inverter always stays well below the gate oxide breakdown voltage. Unlike dev1, dev2 does not need to be very big but it requires a low holding voltage.

For some applications, e.g. for high-frequency pins in RF IC's, the above topology may not be acceptable from a functional perspective. Most common reasons are related to an unacceptable degradation in noise performance, an unacceptable high-frequency behavior, and matching problems. Also the linearity might be overaffected for instance in high frequency (HF) filters. The influence of the input protection for low noise amplifiers in particular will be discussed in Chapter 5.

3.3.2.2 Power Supply Clamping

The aim of the power supply clamp is to provide an explicit discharge path between the power rails as depicted in Fig. 3.15. It may consist of diodes or MOS transistors. For clamps based on MOS transistors the most simple and frequently used structures are based on a grounded gate NMOS or gate coupled NMOS which have been discussed in Section 3.3.1.2 and Section 3.3.1.3 respectively.

The diodes have the advantage that they are very area efficient and easy to implement. Such a diode clamp consists of a series connection of several diodes. The number of diodes depends on the operation voltage of the circuit and the high current resistance requirements. More diodes in the string will reduce the V_{DD} to V_{SS} leakage current but will increase the high current resistance.

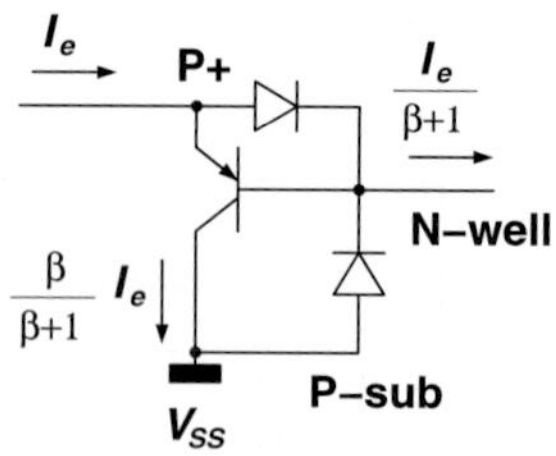

Figure 3.17: *P+ N-well diode as pnp transistor.*

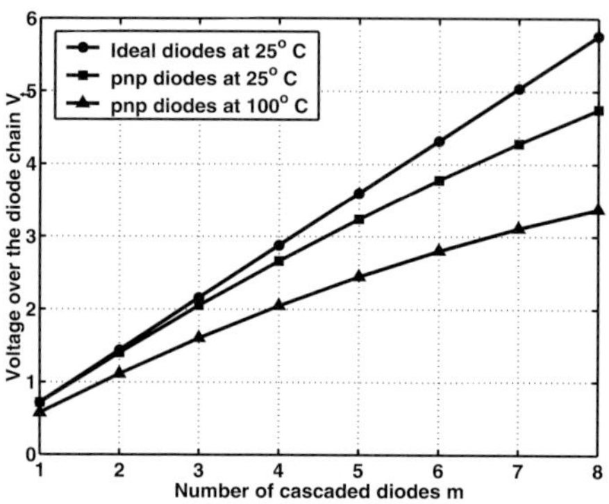

(a) Voltage over the diode chain.

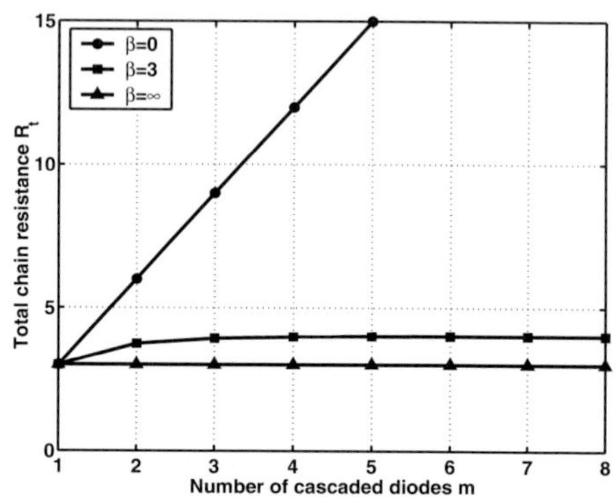

(b) AC resistance of the diode chain.

Figure 3.18: *Behavior of the diode chain as function of the number of diodes in the chain.*

Ideally, it would suffice to divide the supply voltage by the allowed diode voltage to know the number of diodes needed in the clamp. For instance, suppose the operation voltage is 2V. Taking an allowable diode voltage of 0.5V, would make it sufficient to place a string of 4 diodes. The diodes can then be sized according to the required on-resistance.

However, as usual things are a bit more complicated. Since these diodes constitute a forward diode string, their anodes must not be connected to ground but must remain floating. The only suitable diode in a n-well CMOS technology is the p+ n-well diode. Looking at this diode in a bit more detail reveals that it is in fact a pnp transistor [Dab98]. This is illustrated in Fig. 3.16, which shows a chain of four serially connected diodes. This parasitic nature provides the diode with some interesting properties which can be beneficial as well as detrimental.

The single transistor element is shown in Fig. 3.17. It is seen that the CMOS diode is no longer a two-terminal device but has become a three-terminal device. If ideal diodes are cascaded, the cut-in voltage increases linearly with the number of diodes. However, for a diode chain constructed using pnp transistors, the linear increase in cut-in voltage is lost. The pnp action allows some fraction of the emitter current to sink into the substrate; thus there is less current in the next stage. This in turn reduces the voltage over the next stage. This action (identical to

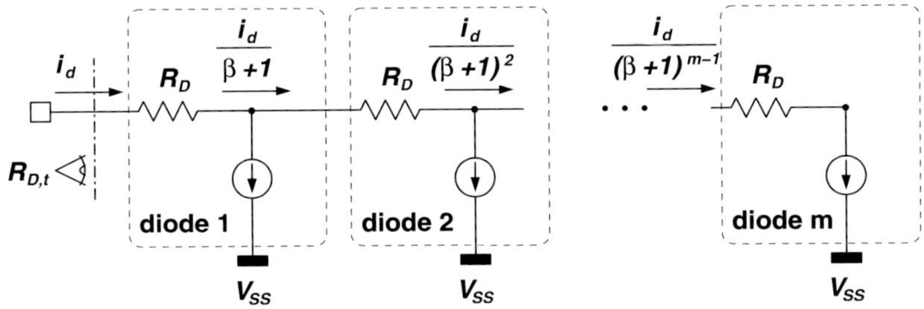

Figure 3.19: *Schematic for calculation of the AC input resistance of the diode clamp.*

a Darlington pair) is repeated in each stage. Thus the sum of the voltages over the cascaded pnp diodes is smaller then it would be for the ideal diodes. The cut-in voltage of identically cascaded diodes is given by

$$\ln\frac{I_1}{I_S} = \frac{qV_1}{nkT} \tag{3.7}$$

$$\ln\frac{I_2}{I_S} = \frac{qV_2}{nkT} = \ln\frac{I_1}{(\beta+1)I_S} = \ln\frac{I_1}{I_S} - \ln(\beta+1), \tag{3.8}$$

where V_1 and I_1 represent the voltage over and current through the top diode, V_2 and I_2 represent the voltage and current for the second diode and so on. The voltage over the second diode can be rewritten as

$$V_2 = V_1 - \frac{nkT}{q}\ln(\beta+1). \tag{3.9}$$

Now let $V_0 = nkT/q$, which is 26mV for an ideal diode at room temperature. The analysis of (3.9) is applied to multiple stages to give a loss of an additional $V_0\ln(\beta+1)$ at each stage, resulting in a total voltage V_t of a string of m identical diodes at current I_1:

$$V_{D,t} = mV_1 - V_0\ln(\beta+1)(\frac{1}{2}m(m-1)). \tag{3.10}$$

This effect is shown graphically in Fig. 3.18(a) for a β of 3. In reality, the quantity β is dependent on the technology and on the specific layout of the diodes. Fig. 3.18(a) plots the voltage over the clamp for a fixed total leakage current set as the maximum allowed leakage current during normal operation. Investigation of this plot allows the designer to choose the number of diodes in the chain based on the maximum supply voltage and the maximum temperature of operation.

It has been stated that cascading ideal diodes increases the cut-in voltage linearly. In that case, also the resistance increases linearly. If the diode chain resistance is to be kept constant, the diode area must be scaled up with the same factor as the number of diodes (m) . Therefore, the total diode chain area scales up with m^2. However since the cut-in voltage increases sublinearly, also the resistance of the forward path is less then expected in an ideal cascaded chain. Once the diode turns on, it takes a low voltage to increase the current through them. This is modeled in Fig. 3.19. The larger voltage drop component then is due to the IR drop in the diode. Assuming there is uniform pnp transistor action in the chain, the input AC resistance $R_{D,t}$ can be calculated by

$$R_{D,t} = R_D \sum_{k=1}^{m} \frac{1}{(\beta + 1)^{k-1}}, \tag{3.11}$$

where R_D is the individual diode resistance. For small β, the resistance of the diode chain would increase close to linearly ($R_{D,t} = mR_D$ for $\beta = 0$ or no pnp action). For large β, the resistance would approach that of a single diode since most of the current would be sunk to ground in the first stage. When β is zero, the diode needs to be scaled in area as m^2 to maintain the same overall resistance. However, in reality, as can be seen in (3.11), it can be scaled to a somewhat smaller degree. Fig. 3.18(b) plots the total AC resistance of the clamp as a function of the number of diodes in the chain for $\beta = 0$, 3 and ∞.

In conclusion, it has been demonstrated that the cascading of the diodes has two effects which should be taken into consideration in the design of a diode based power supply clamp:

- Sublinear increase of the forward cut-in voltages at a constant leakage current.
- Sublinear increase in the total chain resistance.

3.4 Conclusion

This chapter has given a compact overview of the domain of ESD-protection in CMOS. The main ESD tests have been introduced and the differences between them have been investigated. These differences involve the shape and frequency content of the ESD pulses. They have all been related to the implementation of the test setup. The different protection standards are based on these tests. For each test the standards provide several classes of protection, applicable for different products. For IC's, by far the most common standard is the class II HBM standard. The corresponding level of protection is 2 kV. The ESD performance of the chips discussed in Chapter 6 has been measured with TLP and HBM tests. The TLP test is used to give a swift estimate of the ESD performance. The HBM measurement allows to verify this and categorize the chip in the corresponding protection class.

A second part of this chapter was devoted to introducing the most common ESD devices: diodes, GGNMOS, SCR, GCNMOS. Their operation has been discussed based on their physical properties. Different standard ESD protection topologies have been explained both for I/O protection and supply clamping. It has been concluded that the I/O protection circuit presents the core circuit with huge parasitics which may have a serious impact on their performance. Especially in the analog and RF domain where high frequency and low noise performance is required. The parasitics of the supply clamps are usually not important to the core circuit performance. However, even the simple diode string supply clamp presents the designer with some peculiar properties which may be beneficial as well as detrimental. It has been shown that every diode in the string acts as a pnp transistor. This property provides the string with both a sublinear increase in the cut-in voltage which raises the leakage current, and a sublinear increase in the on-resistance which lowers the holding voltage.

Chapter 4

Detailed Study of the Common-Source LNA with Inductive Degeneration

4.1 Introduction

The operation of the CS LNA with inductive degeneration has been explained in Section 2.6.1. A very rudimentary performance model has been introduced. Equations have been developed describing the behavior of the most important performance parameters within the design space of the amplifying transistor. Contour plots have been used to illustrate and clarify the behavior. A comparison was made with the CG LNA and the shunt-feedback LNA. In this chapter, this understanding will be broadened to incorporate the effect of non-idealities.

These non-idealities will be introduced gradually in order to elucidate their particular influence on the LNA performance. The first and very important non-ideality is the presence of the non-quasi static gate resistance. Its influence is investigated in Section 4.2. Another extremely important parasitic is the input capacitance. The influence of this capacitance is of the upmost importance for the input ESD design and will force the designer to seek out alternatives as will be explained in Chapter 5. The role of the parasitic input capacitance is investigated in Section 4.3. The influence of the Miller effect on the performance is explained in Section 4.4. Section 4.5 elaborates on the related optimization of the cascode transistor. The effect of the non-linearity of the output capacitance is discussed in Section 4.6. In Section 4.7 the impact of a finite input match on the overall LNA performance is studied. Several aspects of the load impedance and output matching topologies are considered in Section 4.8. Section 4.9 discusses the LNA bandwidth requirements. The most important layout aspects of integrated low-noise amplifiers are treated in Section 4.10. This chapter concludes in Section 4.11 with a more accurate comparison with the common-gate amplifier in order to predict future trends.

4.2 The Non-Quasi Static Gate Resistance

The non-quasi static gate resistance, $r_{g,NQS}$ was introduced in Section 2.3.3 but neglected in the performance model for the inductively degenerated CS LNA in Section 2.6.1. This is not entirely

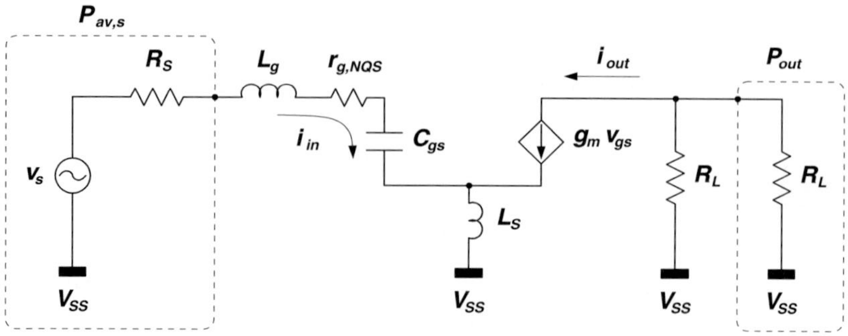

Figure 4.1: *Small signal schematic of the LNA in Fig. 2.17 including the NQS gate resistance.*

—or more accurately, entirely not— justified. Since the presence of $r_{g,NQS}$ results in an extra resistive part in the input impedance, it has a serious influence on the input match calculations. Moreover, the NQS effect also features an extra noise source which should be considered in the noise figure model.

4.2.1 Influence of $r_{g,NQS}$ on Z_{in}, G_T and IIP3

The small-signal schematic of the LNA incorporating the NQS gate resistance is shown in Fig. 4.1. The input impedance is given by

$$Z_{in} = \frac{1}{j\omega C_{gs}} + j\omega(L_g + L_s) + \omega_T L_s + r_{g,NQS} \tag{4.1}$$

which reduces to

$$Z_{in} = \omega_T L_s + r_{g,NQS}, \tag{4.2}$$

at the operating frequency f_0. The resonance requirement has not changed with respect to Section 2.6.1 and is still calculated by (2.91). However the required source inductance L_s is lower and consequently L_g is increased by the same amount to maintain the same frequency of operation:

$$L_s = \frac{R_S - r_{g,NQS}}{\omega_T} \tag{4.3}$$

$$L_g = \frac{1}{\omega_0^2 C_{gs}} - L_s \tag{4.4}$$

Obviously L_s should be positive[1]. Consequently

$$R_S \geq r_{g,NQS} \qquad \text{or} \qquad g_m \geq \frac{1}{\kappa R_S} \tag{4.5}$$

[1]The obvious solution of replacing the inductor by a capacitor does not work since the impedance of the capacitor has a reverse frequency dependency. As a consequence the circuit would become unstable at lower frequencies.

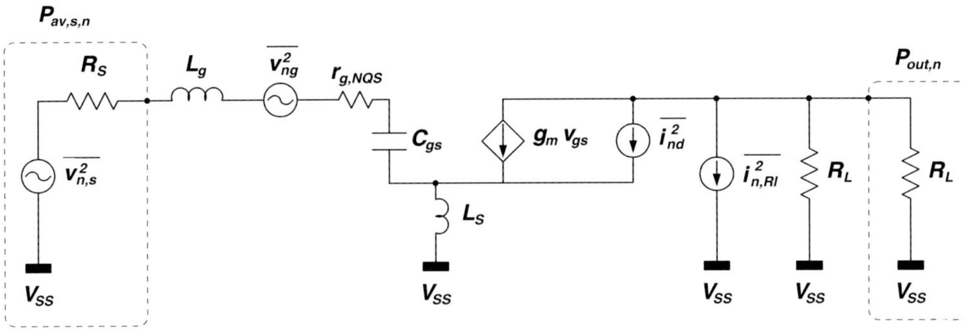

Figure 4.2: *Simplified small signal schematic of the LNA in Fig. 2.17 indicating the main noise sources and including the non-quasi static gate resistance.*

Given these new values for L_g and L_s, the power gain remains unaltered:

$$G_T = \frac{R_L}{4R_S}\left(\frac{\omega_T}{\omega_0}\right)^2. \tag{4.6}$$

Since the voltage across the gate-source capacitance is also unchanged, the IIP3 of the LNA is still given by

$$\text{IIP3} = 5.25 + 10\log\left(\frac{V_{GST}\left(2+\Theta V_{GST}\right)\left(1+\Theta V_{GST}\right)^2}{\Theta}\right) + 20\log\left(2\omega_0 R_S C_{gs}\right). \tag{4.7}$$

4.2.2 Influence of $r_{g,NQS}$ on the Noise Figure

Calculation of the noise figure of the LNA becomes somewhat more complicated due to the introduction of the NQS effect. The main noise sources are shown in Fig. 4.2. Part of the noise of the NQS gate resistance is correlated with the drain noise as expressed in (2.61). Consequently the output noise currents or voltages must be added, not the noise power contributions. Afterwards the output signals can be converted into a total output noise power. The calculation of the noise figure is based on the fundamentals of two-port noise theory explained in Appendix A. The noise factor is calculated by means of the four noise parameters Y_s, R_n, G_u and Y_c. Derivations are omitted but the results are listed in Table 4.1. Substituting these equations in (A.7) yields

$$F \approx 1 + \frac{\frac{\alpha\delta\left(1-|c|^2\right)}{\kappa g_m R_S^2} + \left(\frac{4}{R_S^2} + \frac{|c|^2\alpha^2}{R_S^2}\frac{\delta}{\gamma\kappa}\right)\frac{1}{4}\frac{\gamma}{\alpha}\left(\frac{\omega_0}{\omega_T}\right)^2 g_m\left(R_S + \frac{1}{\kappa g_m}\right)^2}{\frac{1}{R_S}}. \tag{4.8}$$

Since

$$\frac{|c|^2\,\alpha^2\delta}{4\gamma\kappa} \ll 1, \tag{4.9}$$

Noise Parameter	Equation
Y_s	$\frac{1}{R_S}$
R_n	$\frac{1}{4}\frac{\gamma}{\alpha}\left(\frac{\omega}{\omega_T}\right)^2 g_m\left(R_S+\frac{1}{\kappa g_m}\right)^2$
G_u	$\frac{\alpha\delta\left(1-\lvert c\rvert^2\right)}{\kappa g_m R_S^2}$
Y_c	$\frac{1}{R_S}\left(1-c\alpha\sqrt{\frac{\delta}{\gamma\kappa}}\right)$

Table 4.1: *Two-port noise parameters for the CS LNA in Fig. 4.2.*

and

$$\alpha\delta\left(1-\lvert c\rvert^2\right) \gg \frac{\gamma}{\alpha\kappa}\left(\frac{\omega_0}{\omega_T}\right)^2, \tag{4.10}$$

(4.8) can be simplified to

$$F \approx 1+\frac{\gamma}{\alpha}\left(\frac{\omega_0}{\omega_T}\right)^2\left(g_m R_S+\frac{2}{\kappa}\right)+\frac{\alpha\delta\left(1-\lvert c\rvert^2\right)}{\kappa g_m R_S}. \tag{4.11}$$

The noise factor can be split up into two contributions:

$$F = 1+(F_d-1)+(F_g-1). \tag{4.12}$$

The first contribution (F_d-1) stems from the classical drain noise current

$$F_d - 1 \approx \frac{\gamma}{\alpha}\left(\frac{\omega_0}{\omega_T}\right)^2\left(g_m R_S+\frac{2}{\kappa}\right) \tag{4.13}$$

where stating that $\kappa = \infty$ yields the classical drain noise contribution when $r_{g,NQS} = 0$. The same result is then obtained as in Section 2.6.1.3. (F_d-1) is plotted in Fig. 4.3(a). The noise contribution of the classical drain noise current increases towards the upper left corner. This is equivalent to the discussion in Section 2.6.1.3. However, due to the presence of the NQS gate resistance no input match can be obtained in the patterned region bounded by (4.5) and the design space is limited accordingly.

The second contribution (F_g-1) stems from the induced gate noise or more accurately, from the uncorrelated part of the induced gate noise[2]:

$$F_g - 1 = \frac{\alpha\delta\left(1-\lvert c\rvert^2\right)}{\kappa g_m R_S}. \tag{4.14}$$

[2]The correlated part is negligible as shown by (4.9).

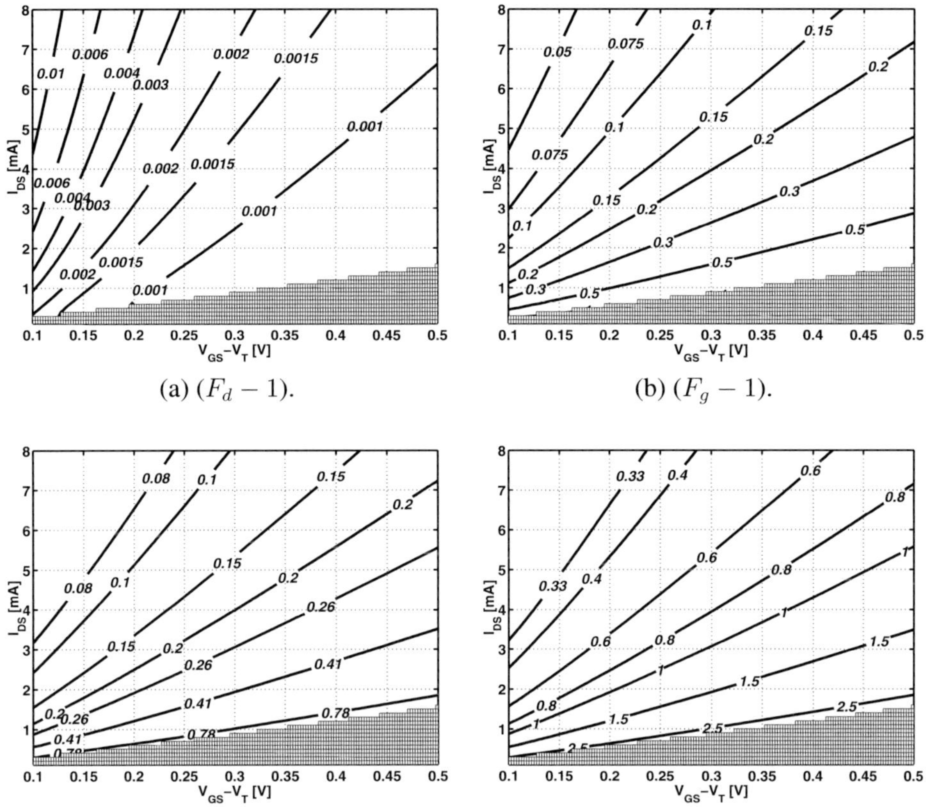

(a) $(F_d - 1)$. (b) $(F_g - 1)$.

(c) $(F - 1)$. (d) $\mathrm{NF} = 10\log(F)$ [dB].

Figure 4.3: *Contours of the different LNA noise factor contributions and the total noise figure.*

Remembering that $r_{g,NQS} = (\kappa g_m)^{-1}$ and that the induced gate noise voltage is given by (2.60), shows that this part of the noise factor simply stems from the ratio of the squared induced gate noise voltage (the uncorrelated part) and the squared source noise voltage:

$$F_g - 1 = \frac{v_{ng}^2(1 - |c|^2)}{v_{n,s}^2}. \tag{4.15}$$

Indeed, Fig. 4.2 shows that both noise sources are simply connected in series. $(F_g - 1)$ is plotted in Fig. 4.3(b). Contrary to the classical drain noise contribution, this contribution increases towards the lower right. Equation (4.14) shows that $(F_g - 1)$ is inversely proportional to g_m which decreases towards the lower right corner.

Considering the behavior of both noise contributions, it becomes clear that the total noise factor will be determined by the classical drain noise when going to the upper left and more by the NQS noise when going to the lower right corner. In this example the transition between both

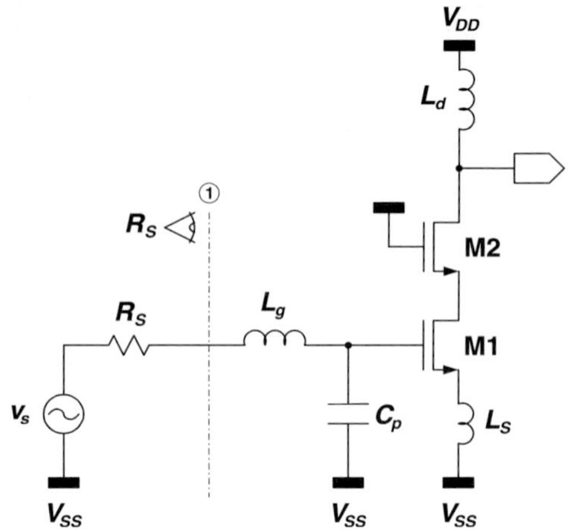

Figure 4.4: *Schematic of the CS LNA including the parasitic input capacitance* C_p.

is off the plot. It is located further to the upper right. Consequently for the complete design subspace shown in the plot the NQS noise contribution is dominant. However this will change once the parasitic input capacitance is taken into account in Section 4.3.

Besides the two already mentioned noise factor contributions, a third one stems from the equivalent load resistance R_L. The same result is obtained as the load contribution $(F_L - 1)$ in (2.101). Indeed, $i_{n,Rl}$ is directly injected in the output node. giving an output noise power of $i_{n,Rl}^2 R_L/4$. Since the power gain is unchanged with respect to Section 2.6.1.2, $(F_L - 1)$ is again given by

$$F_L - 1 = G_T^{-1} = 4\left(\frac{\omega_0}{\omega_T}\right)^2 \frac{R_S}{R_L}. \tag{4.16}$$

The total noise factor F is now found as

$$\begin{aligned} F &= 1 + (F_d - 1) + (F_g - 1) + (F_L - 1) \\ &= 1 + \frac{\gamma}{\alpha}\left(\frac{\omega_0}{\omega_T}\right)^2 \left(g_m R_S + \frac{2}{\kappa}\right) + \frac{\alpha\delta\left(1 - |c|^2\right)}{\kappa g_m R_S} + 4\left(\frac{\omega_0}{\omega_T}\right)^2 \frac{R_S}{R_L}. \end{aligned} \tag{4.17}$$

4.3 Parasitic Input Capacitance

One of the most important non-idealities is the presence of parasitic capacitance at the input of the LNA. The basic schematic of the circuit is shown in Fig. 4.4 where C_p denotes the parasitic input capacitance. This capacitance has four main contributors:

- the input bonding pad,
- the potential input ESD-protection network,

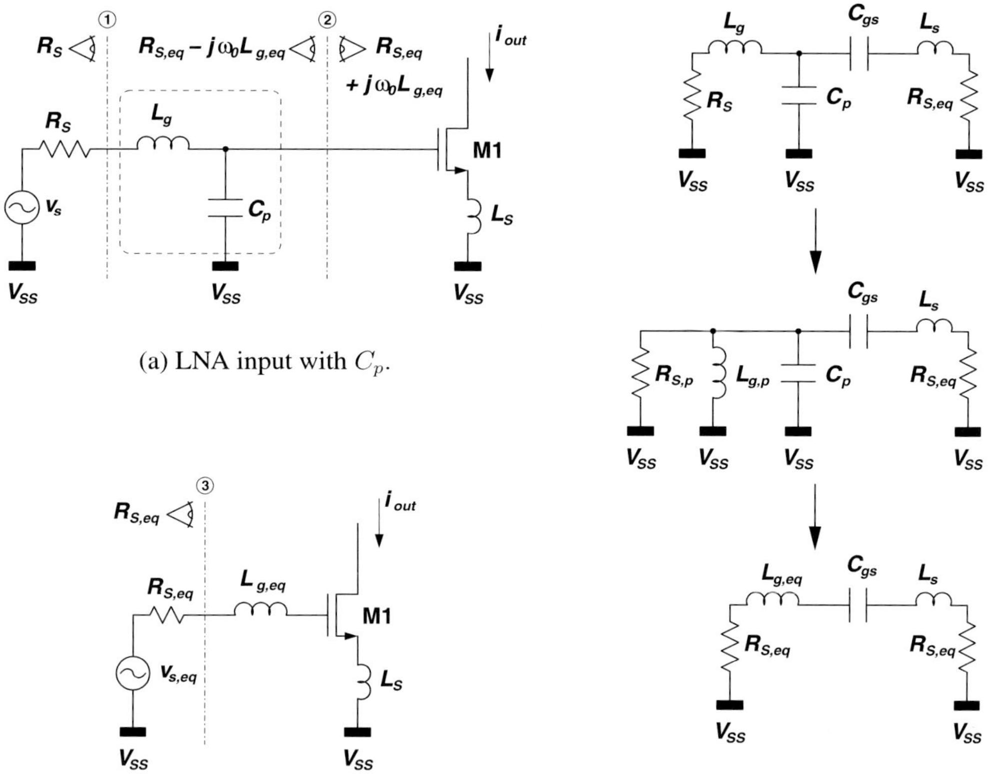

(a) LNA input with C_p.

(b) Equivalent LNA input valid at ω_0.

(c) Two-step transformation of R_S to $R_{S,eq}$ via $R_{S,p}$.

Figure 4.5: *Equivalent schematics of the LNA input.*

- the gate-drain capacitance of the amplifying device,
- the wiring capacitance.

The total capacitance may have a value of 100 fF to more than 1 pF.

4.3.1 Impact of C_p

In this section, the influence of C_p on the performance of the LNA will be evaluated. For the moment it is assumed that C_p behaves 'ideally'. This means it is completely linear and lossless (it has an infinite Q). These constraints will be omitted in Section 4.3.2 and Section 4.3.3.

The input section of the LNA is displayed in Fig. 4.5(a). If C_p were zero then the gate of M1 would see a signal source with a complex source impedance Z_S given by

$$Z_S = R_S + j\omega L_g, \tag{4.18}$$

where $R_S = 50\ \Omega$. However with the presence of C_p, L_g and C_p constitute a lossless L-type transformation network with input at reference plane ① and output at reference plane ②. Consequently, looking to the left of reference plane ② the gate of M1 now sees a different impedance. At the operating frequency of the LNA, this equivalent source impedance can again be split into a real and imaginary part:

$$Z_{S,eq} = R_{S,eq} + j\omega_0 L_{g,eq}, \tag{4.19}$$

where $R_{S,eq}$ and $L_{g,eq}$ are the resistive and inductive part of the new, equivalent source impedance as indicated in Fig. 4.5(b). They can be found through the two-step series-parallel transformation shown in Fig. 4.5(c), valid at ω_0. $R_{S,p}$ is the equivalent parallel source resistance given by

$$R_{S,p} = \frac{\omega_0^2 L_g^2}{R_S} + R_S = \frac{\omega_0^2 L_{g,eq}^2}{R_{S,eq}} + R_{S,eq}. \tag{4.20}$$

Both $R_{S,eq}$ and $L_{g,eq}$ can be calculated as a function of C_p, L_g, R_S and ω_0:

$$R_{S,eq} = \frac{R_S}{\omega_0^2 C_p^2 R_S^2 + (1 - \omega_0^2 C_p L_g)^2} \tag{4.21}$$

$$L_{g,eq} = \frac{L_g - C_p\left(\omega_0^2 L_g^2 + R_S^2\right)}{\omega_0^2 C_p^2 R_S^2 + (1 - \omega_0^2 C_p L_g)^2}. \tag{4.22}$$

Theoretically, $L_{g,eq}$ can become negative but this possibility will drop out as soon as the input matching criterium is introduced.

Since the presence of C_p has now been reformulated into an equivalent source resistance $R_{S,eq}$ and gate inductor $L_{g,eq}$, the analysis in Section 2.6.1 and Section 4.2 can be reused to visualize the resulting performance.

4.3.1.1 Influence of C_p on Input Matching

Referring to Fig. 4.5, the input matching criterium should be rewritten. The equivalent input impedance seen to the right of reference plane ③ is given by

$$Z_{in,eq} = \frac{1}{j\omega C_{gs}} + j\omega(L_{g,eq} + L_s) + \omega_T L_s + r_{g,NQS}. \tag{4.23}$$

The complex matching equation $Z_{in,eq} = R_{S,eq}$ at $\omega = \omega_0$ together with (4.21) and (4.22) allows the calculation of $L_{g,eq}$, L_g, L_s and $R_{S,eq}$ as a function of M1 parameters and the value of C_p. Assuming that $\omega_T^2 \gg \omega_0^2$, $R_{S,eq}$ is approximated by

$$R_{S,eq} \approx \frac{2R_S\left(1 + \frac{C_p\omega_T}{g_m}\right)^2}{\psi + \sqrt{\psi^2 - (\psi - 1)^2\left(\frac{\omega_T}{\omega_0}\right)^2}}, \tag{4.24}$$

where

$$\psi = 1 + \frac{2C_p R_S \omega_0^2}{\omega_T}\left(1 + \frac{C_p\omega_T}{g_m}\right). \tag{4.25}$$

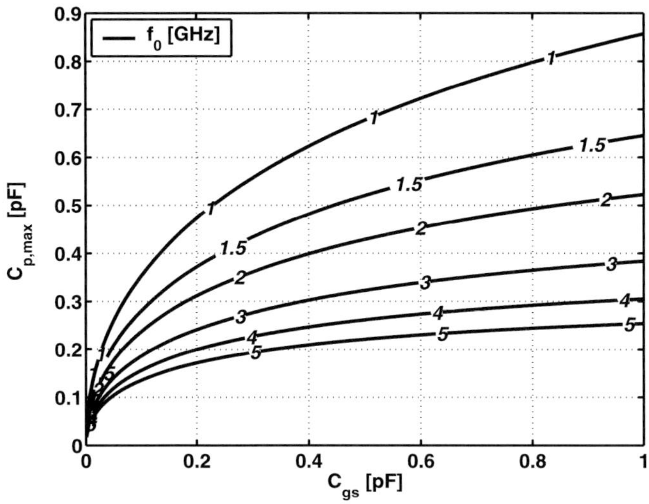

Figure 4.6: *Maximum allowed value of C_p for input matching.*

Consequently, taking the input matching constraint into account, the equivalent source resistance is shown to increase for increasing C_p but its exact value depends on the design of transistor M1. The design space of M1 is still limited by (4.5) which can be rewritten as

$$g_m \geq \frac{1}{\kappa R_{S,eq}}. \tag{4.26}$$

However, another, more stringent constraint has arisen. Namely $R_{S,eq}$ needs to be a *real* number. Consequently

$$\psi^2 - (\psi - 1)^2 \left(\frac{\omega_T}{\omega_0}\right)^2 \geq 0, \tag{4.27}$$

or

$$R_S \leq \frac{\omega_T}{2C_p\omega_0^2\left(\frac{\omega_T}{\omega_0} - 1\right)\left(1 + \frac{C_p\omega_T}{g_m}\right)}. \tag{4.28}$$

Assuming that $\omega_T \gg \omega_0$ this requirement reduces to

$$R_S \leq \frac{1}{2C_p\omega_0\left(1 + \frac{C_p}{C_{gs}}\right)}. \tag{4.29}$$

If above equation is not satisfied, then no input match can be obtained for the given combination of M1 and C_p. Clearly matching becomes more difficult when C_p becomes larger. The maximum value of C_p for matching is found from (4.29):

$$C_{p,max} = \frac{1}{2}\left(-C_{gs} + \sqrt{C_{gs}^2 + \frac{2C_{gs}}{R_S\omega_0}}\right). \tag{4.30}$$

It is plotted versus C_{gs} in Fig. 4.6 for frequencies ranging from 1 to 5 GHz. Already intuitively one can see here that larger C_p values will generally lead to a larger M1 width. It is also seen from (4.29) and Fig. 4.6 that larger ω_0 will severely restrict the M1 design space.

4.3.1.2 Influence of C_p on Power Gain, Noise Figure and IIP3

Since the L_g-C_p transformation network is supposed lossless, the available power of the equivalent signal source ($v_{s,eq}$ and $R_{S,eq}$) has remained unaltered:

$$P_{av,s} = \frac{v_s^2}{4R_S} = \frac{v_{s,eq}^2}{4R_{S,eq}}. \tag{4.31}$$

The power gain is found by replacing R_S by $R_{S,eq}$ in (2.98).

$$G_T = \frac{R_L}{4R_{S,eq}}\left(\frac{\omega_T}{\omega_0}\right)^2. \tag{4.32}$$

For any given point in the design space of M1, $R_{S,eq}$ becomes larger for increasing C_p. Hence, the corresponding power gain drops. Another, more intuitive way of looking at this phenomenon is that, since more signal current is sunk into C_p less current can flow through C_{gs} and can be used to generate output current. This means the input current is used less efficiently to generate output current. The power gain is plotted in Fig. 4.7(d) for a C_p of 210 fF [3]. The contribution of C_{gd} is neglected for now. Its influence is discussed in Section 4.4. The corresponding $R_{S,eq}$ and Q_{in} are shown in Fig. 4.7(b) and Fig. 4.7(c) respectively. Q_{in} was defined by (2.100) and is recalculated as

$$Q_{in} = \frac{v_{gs}}{v_s} = \frac{1}{2\omega_0 C_{gs}\sqrt{R_{S,eq}R_S}}. \tag{4.33}$$

The noise figure of the LNA is described in a very similar way. Equation (4.17) is reused where R_S is replaced by $R_{S,eq}$:

$$\begin{aligned} F &= 1 + (F_d - 1) + (F_g - 1) + (F_L - 1) \\ &= 1 + \frac{\gamma}{\alpha}\left(\frac{\omega_0}{\omega_T}\right)^2\left(g_m R_{S,eq} + \frac{2}{\kappa}\right) + \frac{\alpha\delta\left(1-|c|^2\right)}{\kappa g_m R_{S,eq}} + 4\left(\frac{\omega_0}{\omega_T}\right)^2\frac{R_{S,eq}}{R_L}. \end{aligned} \tag{4.34}$$

Even though the noise figure is clearly influenced by the value of $R_{S,eq}$ and hence C_p, the resulting performance is not necessarily inferior. The classical drain noise contribution ($F_d - 1$) increases due to the lower Q-factor of the input stage. The contribution of the load resistance also increases due to the lower power gain. However the contribution of the NQS-gate resistance is lowered due to its decreased relative significance in the total input impedance. Remember that this noise voltage is directly compared to the input noise voltage which is proportional to $R_{S,eq}$. The total noise factor of the LNA with a C_p of 210 fF is shown in Fig. 4.7(c). It should be compared with Fig. 4.3(d). This comparison shows that the noise figure is indeed lower for

[3] The value of 210 fF was used because this is the actual value present in the design discussed in Section 6.2

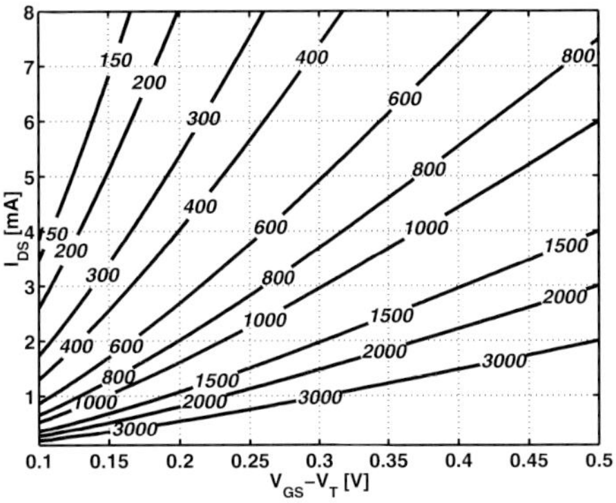

(a) $R_{S,opt}$ [Ω].

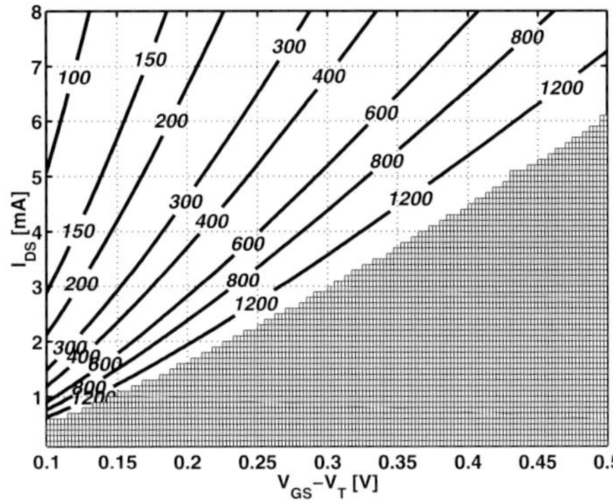

(b) $R_{S,eq}$ [Ω].

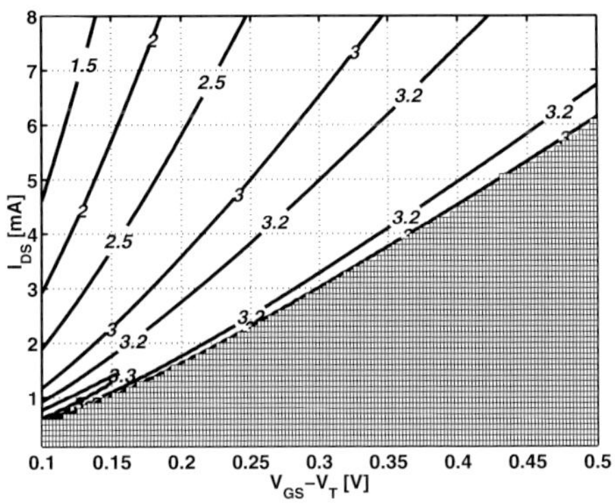

(c) Q_{in} [ ].

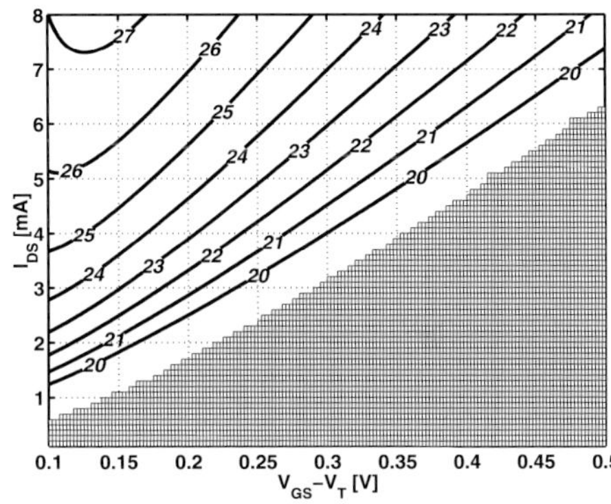

(d) G_T [dB].

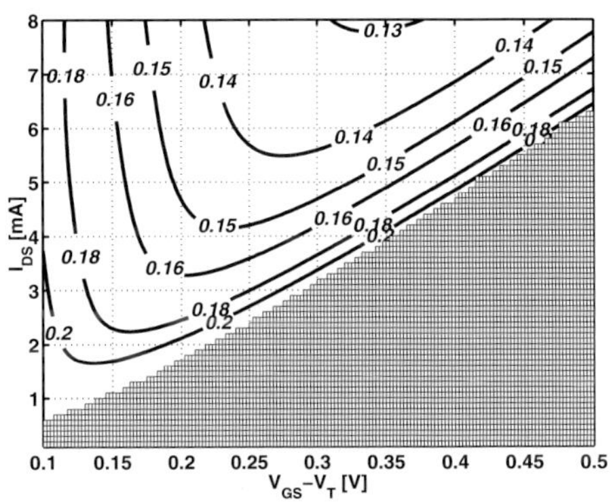

(e) NF [dB].

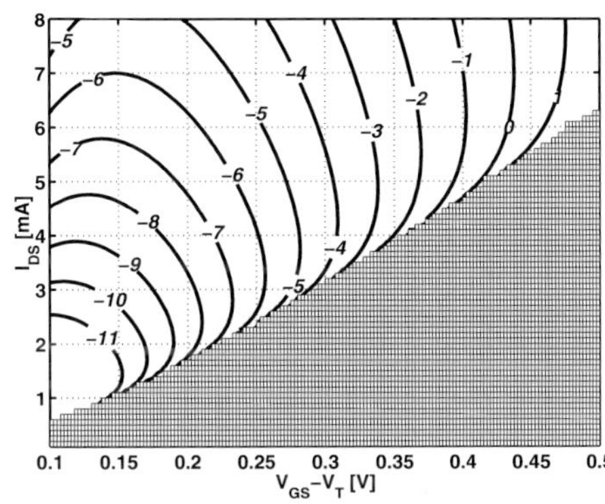

(f) IIP3_{M1} [dBm].

Figure 4.7: *Contour plots of the LNA behavior for a* C_p *of 210 fF.*

a large part of the design space. In Fig. 4.3(d) no C_p was present and in most part of the design space (except the upper left corner), the NQS noise was dominant. However in Fig. 4.7(c), the classical noise dominates throughout the entire M1 design space.

It is interesting to see what value of $R_{S,eq}$ minimizes the noise figure in any point of the design space. This optimum source resistance will be called $R_{S,opt}$. It is found by differentiating (4.34) with respect to $R_{S,eq}$ and equating the result to zero. This results in

$$R_{S,opt} = \left(\frac{\omega_T}{\omega_0}\right)\sqrt{\frac{\alpha\delta\left(1-|c|^2\right)}{\kappa g_m\left(\frac{\gamma}{\alpha}g_m + \frac{4}{R_L}\right)}}. \tag{4.35}$$

$R_{S,opt}$ is depicted in Fig. 4.7(a). Comparing Fig. 4.7(a) with Fig. 4.7(b) shows that $R_{S,eq}$ is quite close to $R_{S,opt}$. This explains why the noise figure is in fact lower than without any C_p. However since $R_{S,eq}$ is already higher than $R_{S,opt}$ any additional capacitance will inevitably increase the noise figure. The noise factor corresponding to $R_{S,opt}$ is:

$$F_{opt} = 1 + \frac{2\gamma}{\kappa\alpha}\left(\frac{\omega_0}{\omega_T}\right)^2 + \left(\frac{\omega_0}{\omega_T}\right)\sqrt{\frac{\alpha\delta\left(1-|c|^2\right)\left(\frac{\gamma}{\alpha}g_m + \frac{4}{R_L}\right)}{\kappa g_m}}. \tag{4.36}$$

This shows that if $g_m R_L \gg \frac{4\alpha}{\gamma}$, which is usually fulfilled, that F_{opt} is only dependent on ω_T and hence on $V_{GS} - V_T$ and *not* on the current. Note also that setting $R_{S,eq} = R_{S,opt}$ is different from a noise match. In the latter the complex source impedance is changed to yield the minimum noise figure for a given circuit (cf. (A.9) in Appendix A). In our case the input impedance match is taken as a design constraint and the equivalent (transformed) source impedance is modified together with the source degeneration inductor L_s in order to minimize the noise figure for a given design point of transistor M1. If the resulting circuit was fixed and the impedance matching constraint released, then a real noise match would yield an even lower noise factor (F_{min}) but the amplifier would no longer be matched at the input.

Another noise figure optimization which may be even more interesting is done assuming the power consumption is limited by a maximum value. This is logically called the power constrained optimization. In that case it makes no sense to allow variation of $R_{S,eq}$ since if $R_{S,eq} = R_{S,opt}$ then the noise figure keeps decreasing for larger $V_{GS} - V_T$. Often however $R_{S,eq}$ can not be chosen freely due to other parasitic effects as for instance in Section 5.3 and Section 5.4. Equation (4.34) shows that $F_g - 1$ increases more or less linearly with $V_{GS} - V_T$ for a fixed current. All other contributions decrease for large $V_{GS} - V_T$ due to the higher cut-off frequency. Consequently an optimum $V_{GS} - V_T$ can be found for a fixed current and fixed $R_{S,eq}$. Even though the analytical expression for $(V_{GS} - V_T)_{opt}$ is very complicated, one can easily deduce the behavior of the optimum with respect to the main parameters. If the increase of a certain parameter either increases $F_g - 1$ or decreases $F_d - 1 + F_L - 1$ for the same $V_{GS} - V_T$, then $(V_{GS} - V_T)_{opt}$ will go up. Consequently, $(V_{GS} - V_T)_{opt}$ will increase

- for larger frequencies at the same current and source resistance, since ω_T is always scaled by ω_0 which will increase $F_d - 1$ and $F_L - 1$.

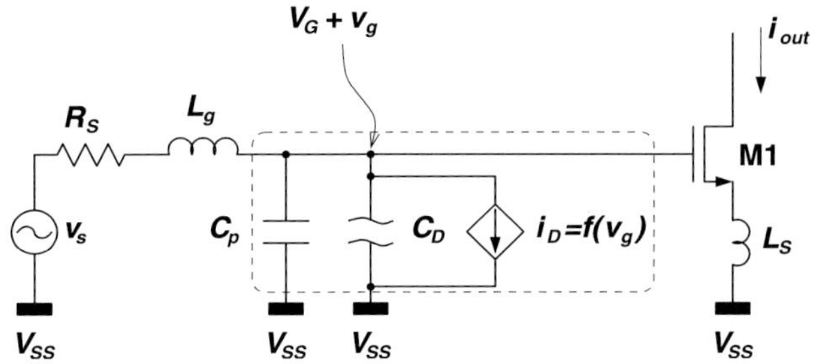

Figure 4.8: *Schematic of the LNA input indicating both linear and non-linear contributions to the parasitic input capacitance* C_p.

- for a larger bias current at the same frequency and source resistance, since $F_g - 1$ is inversely proportional to the current and will be smaller for the same $V_{GS} - V_T$.
- for larger source resistance $R_{S,eq}$ at the same frequency and bias current, since this will both increase $F_d - 1$ and $F_L - 1$ as well as decrease $F_g - 1$.

The impact of C_p on the linearity of the LNA may be described also by considering the changed source resistance. Equation (4.7) is reused yielding

$$\mathrm{IIP3}_{M1} = 5.25 + 10\log\left(\frac{V_{GST}\left(2+\Theta V_{GST}\right)\left(1+\Theta V_{GST}\right)^2}{\Theta}\right) - 20\log\left(Q_{in}\right), \tag{4.37}$$

where Q_{in} is given by (4.33). Comparing (4.32) with (4.37) shows the increase in IIP3 is equal to the reduction in power gain. Both effects result from the decreased value of Q_{in}. Consequently the product of both (OIP3 $=$ IIP3 $\times$ G_T has remained constant. Indeed, it was shown by (2.113) that OIP3 is independent of $R_{S,eq}$. Note that the symbol $\mathrm{IIP3}_{M1}$ was used here instead of IIP3 to contrast with $\mathrm{IIP3}_p$ which refers to the IIP3 solely resulting from the C_p non-linearity and with $\mathrm{IIP3}_{db2}$ which refers to the IIP3 solely resulting from the non-linearity of the drain-bulk capacitance of M2, discussed in Section 4.6. The non-linearity of C_p is discussed next.

4.3.2 Impact of C_p Non-Linearity

Up until now the parasitic input capacitance C_p was assumed to be linear. However, if the input of the LNA has an ESD-protection network, then part of C_p can be non-linear, for instance because it contains a diode capacitance. This non-linearity needs to be considered for the global linearity of the LNA.

In order to investigate the importance of this non-linearity, the IIP3 of the LNA is recalculated taking only the non-linearity of the input capacitance into account. The input schematic of the LNA is shown in Fig. 4.8. For the numerical evaluations it is assumed that C_p is again 210 fF and that part of this belongs to the junction capacitance C_{D0} of a diode which can be a part of the input ESD-protection. It is discussed further in Section 5.2. The fraction of C_{D0} in the

Diode parameter	Value
C_J [F/m^2]	0.0017
V_{bi} [V]	1.1
M_J [ ]	0.57
V_G [V]	0.7

Table 4.2: *Bottom-plate junction parameters of the p+ n-well diode.*

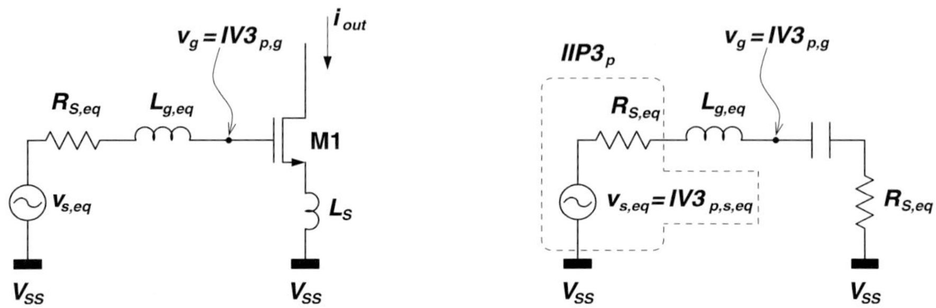

(a) Equivalent input schematic at ω_0. (b) Corresponding small signal schematic.

Figure 4.9: *Schematic of the LNA input for referring the IIP3 voltage of the diode back to the corresponding available power at the input.*

total parasitic input capacitance is denoted ξ_D. Taking only the bottom plate of the junction into account, this capacitance (at biasing conditions) is given by

$$C_{D0} = A_D C_J \left(1 + \frac{V_G}{V_{bi}}\right)^{-M_J} = \xi_D C_p, \tag{4.38}$$

where A_D is the area of the bottom plate of the diode and C_J, V_{bi} and M_J correspond to the SPICE parameters for the junction capacitance per unit area (CJ), the built-in voltage of the diode (PB) and the corresponding exponent (MJ), and V_G is the DC voltage at the gate of M1. The numerical values for the calculations are listed in Table 4.2. The charge Q_D on this capacitance is calculated by

$$Q_D = C_D V_G, \tag{4.39}$$

and can be expanded in a Taylor series around V_G, analogously to the derivations of Section 2.2.5.

The $\mathrm{IV3}_{p,g}$ (voltage amplitude at the gate at intercept) for the stand-alone diode is found by applying (2.36). This yields:

$$\mathrm{IV3}_{p,g} = \sqrt{\frac{4}{3}\left|\frac{(-M_J V_G + V_{bi} + V_G)(V_{bi} + V_G)^2}{M_J(-M_J^2 V_G + 3M_J V_{bi} + V_G + 3V_{bi})}\right|}. \tag{4.40}$$

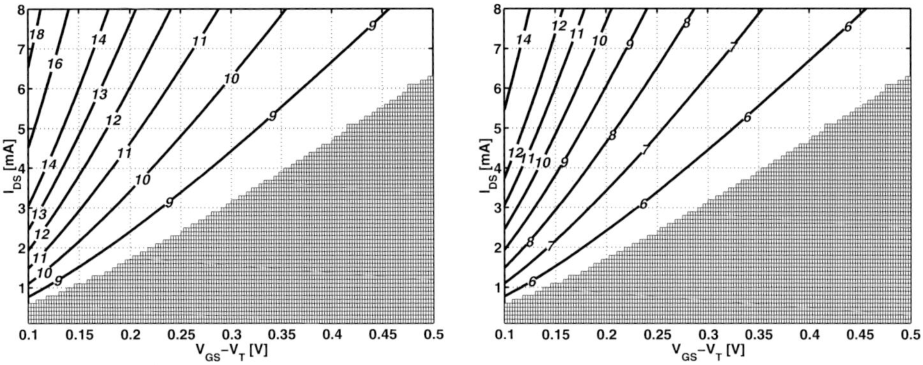

(a) $\mathrm{IIP3}_p$ [dBm] for $\xi_D = 0.25$ or $C_{D0} \approx 50$ fF.

(b) $\mathrm{IIP3}_p$ [dBm] for $\xi_D = 0.5$ or $C_{D0} \approx 100$ fF.

Figure 4.10: *Contour plots of* $\mathrm{IIP3}_p$*, the* $\mathrm{IIP3}$ *solely due to* C_p *non-linearity.*

This voltage amplitude still needs to be converted into the corresponding available source power. This is illustrated by Fig. 4.9. An intermediate conversion yields the corresponding RMS voltage, $\mathrm{IV3}_{p,S,eq}$ of the equivalent voltage source:

$$\mathrm{IV3}_{p,S,eq} = \frac{\mathrm{IV3}_{p,g}}{\sqrt{\xi_D\left(\frac{1}{2} + \frac{\omega_0^2 L_{g,eq}^2}{2R_{S,eq}^2}\right)}}. \tag{4.41}$$

This is converted into available source power and expressed in [dBm]:

$$\begin{aligned}\mathrm{IIP3}_p &= 30 + 10\log\left(\frac{\mathrm{IV3}_{p,S,eq}^2}{4R_{S,eq}}\right)\\ &= 30 + 10\log\left|\frac{4\left(-M_J V_{d0} + V_{bi} + V_{d0}\right)\left(V_{bi} + V_{d0}\right)^2}{3M_J\left(-M_J^2 V_{d0} + 3M_J V_{bi} + V_{d0} + 3V_{bi}\right)}\right|\\ &\quad - 10\log\left(2R_{S,eq} + \frac{2\omega_0^2 L_{g,eq}^2}{R_{S,eq}}\right) - 10\log\left(\xi_D\right).\end{aligned} \tag{4.42}$$

$\mathrm{IIP3}_p$ is plotted in Fig. 4.10 for two different diode fraction, $\xi_D = 0.25$ and $\xi_D = 0.5$, corresponding to a junction capacitance of about 50 fF and 100 fF. Comparing both figures with Fig. 4.7(f) shows that the non-linearity of C_p can be neglected at this frequency (1.57 GHz). The reduced Q_{in} for the same design point at increased frequency will increase $\mathrm{IIP3}_{M1}$. Hence, both contributions will tend to converge. Nevertheless, also at higher frequencies, $\mathrm{IIP3}_{M1}$ will stay dominant.

Since the IIP3 in Fig. 4.10 and Fig. 4.7(f) is expressed in [dBm], the global IIP3 is approximated by

$$\mathrm{IIP3} \approx \min\left(\mathrm{IIP3}_{M1}, \mathrm{IIP3}_p\right). \tag{4.43}$$

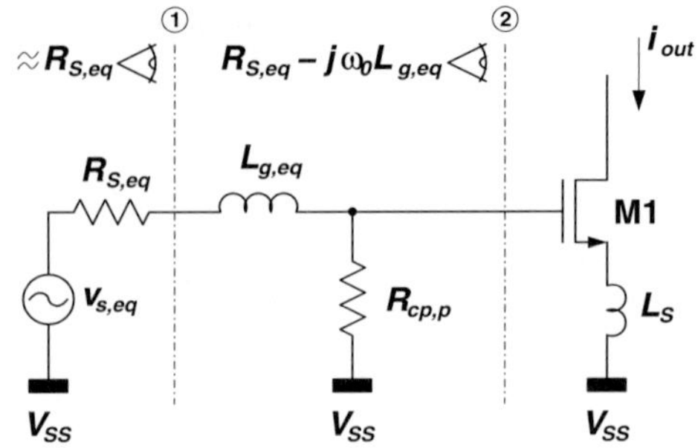

Figure 4.11: *Schematic of the LNA input indicating the influence of the finite Q-factor of* C_p *which results in a finite equivalent parallel resistor* $R_{cp,p}$.

Care should be taken however since above expression gives an upper bound. This can be quite an overestimation if both contributions are close. If so, one can calculate the exact overall IIP3 by

$$\frac{1}{\text{IIP3}} = \frac{1}{\text{IIP3}_{M1}} + \frac{1}{\text{IIP3}_p}, \tag{4.44}$$

where IIP3, IIP3_{M1} and IIP3_p are expressed in [mW], not [dBm].

4.3.3 Impact of the Finite Q of C_p

Another non-ideality of C_p is the finite Q-factor. This means, a non-zero resistance $R_{cp,s}$ is found in series with C_p:

$$Q_{cp} = \frac{1}{\omega_0 C_p R_{cp,s}}. \tag{4.45}$$

This resistance can come from the bonding pad contribution and from the ESD-protection. Naturally, the design and layout are focussed at minimizing this resistance since it adds noise to the circuit and it consumes signal power. The contribution from the bonding pad can be minimized by using a metal shield as explained in Section 4.10.1. The contribution of the ESD-protection should also be minimized, both from RF as from ESD standpoint. Consequently, the resistance is usually rather low and in the order of a few Ohm. Moreover since the resistance is inversely proportional to the area and C_p itself is proportional to the area, the resulting time constant is relatively independent of the actual area or value of C_p. A value of $10\ \Omega$ for a capacitance of 100 fF is surely an overestimation:

$$\tau_{cp} = C_p R_{cp,s} \sim 10^{-12} \quad \text{or} \quad \omega_{cp} = \frac{1}{\tau_{cp}} \sim 10^{12}. \tag{4.46}$$

So in fact we are talking about frequencies in the range of several 100 GHz. The resulting Q-factor for $f_0 = 1.5$ GHz is in the order of 10^2:

$$Q_{cp} = \frac{\omega_{cp}}{\omega_0} \sim 10^2. \tag{4.47}$$

For $C_p = 210$ fF, $R_{cp,s}$ can be converted into an equivalent parallel resistance $R_{cp,p}$:

$$R_{cp,p} \approx Q_{cp}^2 R_{cp,s} \sim 5 \times 10^4\ \Omega. \tag{4.48}$$

This is illustrated in Fig. 4.11. $R_{cp,p}$ can be neglected in the input matching conditions if it is much larger than the equivalent parallel source resistance $R_{S,p}$ defined by (4.20) and depicted in Fig. 4.5:

$$R_{cp,p} \gg R_{S,p} = \frac{\omega_0^2 L_{g,eq}^2}{R_{S,eq}} + R_{S,eq}. \tag{4.49}$$

$R_{S,p}$ is typically in the order of 1 kΩ and hence the condition of equation (4.49) is fulfilled even with the very cautious estimate of (4.46). The input matching conditions of the LNA are unchanged and the power gain remains unaffected. The main change in the noise figure is an extra term added to the noise factor:

$$F_{rcp} - 1 \approx \frac{R_{S,p}}{R_{cp,p}}. \tag{4.50}$$

This shows that as $R_{cp,p}$ decreases (lower Q_{cp}), the noise figure will be the first to change noticeably. Naturally, the goal is to keep this noise contribution as low as possible.

At higher frequencies the influence of Q_{cp} will gradually increase. $R_{cp,p}$ decreases fast with the square of the operating frequency. At 5 GHz, $R_{cp,p}$ will be about ten times lower. Consequently for frequencies beyond 5 GHz, minimizing $R_{cp,s}$ will be of the upmost importance.

4.4 Miller Capacitance

The influence of the gate-drain capacitance of M1 may be split up into four effects. The first effect is the capacitive loading on the input node. This is described by incorporating C_{gd} in the value of C_p. This was discussed in Section 4.3 where C_{gd} was taken as an extra contribution to the fixed part of C_p (210 fF) depending on the actual point in the design space of M1. The second effect refers to the capacitive loading at the drain of M1. This is neglected for now. The third effect represents the feedforward current through C_{gd}. This results in the well known zero at frequency g_m/C_{gd} and can be neglected all together.

The fourth effect is the feedback current through C_{gd} and is known as the Miller effect. The Miller factor M is defined by

$$M = \frac{v_c}{v_{gs}} = \frac{g_m}{g_{m2} + g_{mb2}}(1 + \Lambda I_{DS} R_L) \approx \frac{\alpha g_m}{g_{m2}} \tag{4.51}$$

where g_{m2} and g_{mb2} are the transconductance and bulk transconductance of cascode transistor M2. It was shown in [Jan01] that the Miller effect for M1 can be incorporated in the LNA behavior by reducing the impedance seen to the right of reference plane ① in Fig. 4.12 by a factor $(1 + M\alpha_{gd})$ where

$$\alpha_{gd} = \frac{C_{gd}}{C_{gs}}. \tag{4.52}$$

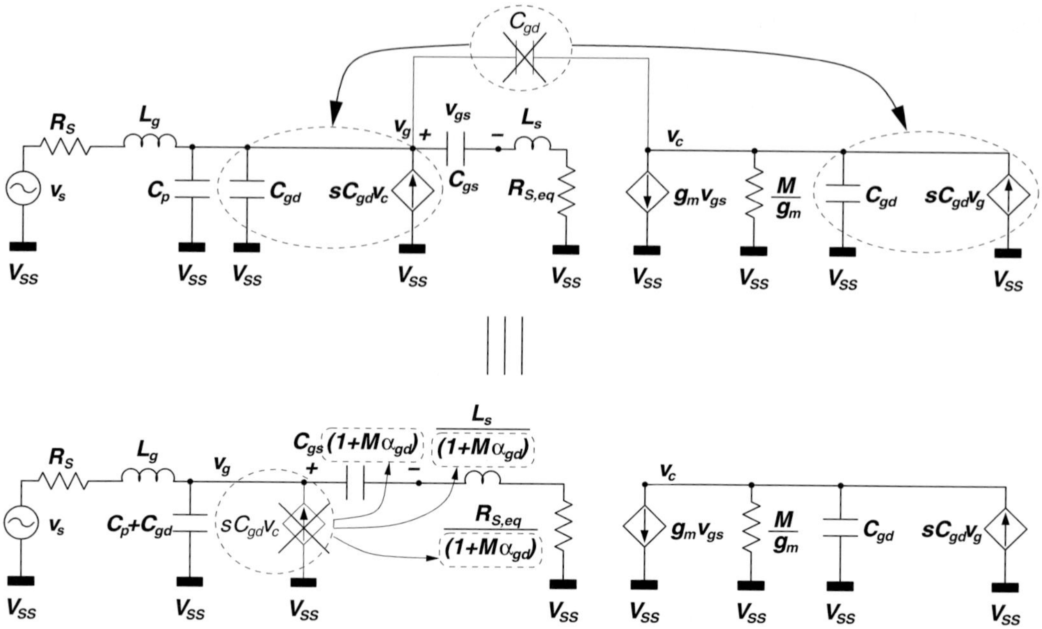

Figure 4.12: *Two-step transformation of C_{gd}. C_{gd} is split up in its Y parameter equivalent. The feedback component can be omitted by replacing the different components as indicated.*

The new value for $R_{S,eq}$ and ψ is found by replacing C_{gs} with $C_{gs}(1 + M\alpha_{gd})$ in (4.24) and (4.25):

$$R_{S,eq} \approx \frac{2R_S\left(1 + \frac{C_p\omega_T}{g_m(1+M\alpha_{gd})}\right)^2}{\psi + \sqrt{\psi^2 - (\psi - 1)^2\left(\frac{\omega_T}{\omega_0(1+M\alpha_{gd})}\right)^2}}, \tag{4.53}$$

$$\psi = 1 + \frac{2C_pR_S\omega_0^2(1 + M\alpha_{gd})}{\omega_T}\left(1 + \frac{C_p\omega_T}{g_m(1 + M\alpha_{gd})}\right). \tag{4.54}$$

In fact this boils down to a mere reduction of ω_T with $(1 + M\alpha_{gd})$. Consequently the feedback effect of C_{gd} yields a decrease in $R_{S,eq}$ whereas the input loading of C_{gd} increased the value of $R_{S,eq}$. The net effect of C_{gd} on $R_{S,eq}$ depends on the actual point of the design space.

G_T, IIP3 and F can be found by replacing all instances of C_{gs} with $C_{gs}(1 + M\alpha_{gd})$ and ω_T with $\omega_T/(1 + M\alpha_{gd})$. The power gain is rewritten as

$$G_T = \frac{R_L}{4R_{S,eq}}\left(\frac{\omega_T}{\omega_0}\right)^2\frac{1}{(1 + M\alpha_{gd})^2}. \tag{4.55}$$

The IIP3 is increased by the same amount as the decrease in gain since both originate from the lower signal current through C_{gs}:

$$\mathrm{IIP3}_{M1} = 5.25 + 10\log\left(\frac{V_{GST}\left(2+\Theta V_{GST}\right)\left(1+\Theta V_{GST}\right)^2}{\Theta}\right) - 20\log\left(Q_{in}\right), \tag{4.56}$$

where

$$Q_{in} = \frac{v_{gs}}{v_s} = \frac{1}{2\omega_0 C_{gs} R_{S,eq}\left(1+M\alpha_{gd}\right)}\sqrt{\frac{R_{S,eq}}{R_S}}. \tag{4.57}$$

The noise factor is found as

$$\begin{aligned} F &= 1 + (F_d - 1) + (F_g - 1) + (F_L - 1) \\ &\approx 1 + \frac{\gamma\left(1+M\alpha_{gd}\right)^2}{\alpha}\left(\frac{\omega_0}{\omega_T}\right)^2\left(g_m R_{S,eq} + \frac{2}{\kappa}\right) + \frac{\alpha\delta\left(1-|c|^2\right)}{\kappa g_m R_{S,eq}} + G_T^{-1}, \end{aligned} \tag{4.58}$$

where G_T is given by (4.55).

4.5 Optimization of the Cascode Transistor

The influence of the cascode transistor on the LNA performance cannot be neglected. The size of M2 determines the Miller-factor M which is present in all design equations (4.55) to (4.58). Moreover M2 will add a noise contribution of its own, $(F_c - 1)$. Since M2 has a large impedance $Z_{s,c}$ connected to the source, the actual noise current at the output is reduced by $|1 + g_{m2}Z_{s,c}|$ where the impedance $Z_{s,c}$ is mostly capacitive and given by

$$Z_{s,c} = \frac{1}{j\omega_0\left(C_{gs}\left(\alpha_{gd}+\alpha_{db}\right) + C_{gs2}\left(1+\alpha_{sb}\right)\right)}. \tag{4.59}$$

This assumption leads to the following equation for $(F_c - 1)$:

$$F_c - 1 = 4\gamma\left(\frac{\omega_0}{\omega_c}\right)^2\left(\frac{\omega_0}{\omega_T}\right)^2 g_m R_{S,eq}\frac{\left(1+M\alpha_{gd}\right)^2}{M}\left(1+\left(\frac{\omega_0}{\omega_c}\right)^2\right), \tag{4.60}$$

where the presence of $Z_{s,c}$ was reformulated as a function of the pole at the cascode node, ω_c:

$$\omega_c = \frac{g_{m2}}{\alpha\left(C_{gs}\left(\alpha_{gd}+\alpha_{db}\right) + C_{gs,2}\left(1+\alpha_{sb}\right)\right)}, \tag{4.61}$$

where α was defined by (2.42). If ω_0 is relatively close to ω_c also the power gain of the LNA will be affected due to the loss of signal current at the cascode node. The gain is reduced to

$$G_T = \frac{R_L}{4R_{S,eq}}\left(\frac{\omega_T}{\omega_0}\right)^2\frac{1}{\left(1+M\alpha_{gd}\right)^2}\frac{1}{1+\left(\frac{\omega_0}{\omega_c}\right)^2}. \tag{4.62}$$

The linearity is unaffected since the signal loss occurs after the non-linearity.

Consequently, the power gain and noise figure depend on the design of the cascode transistor through both M and ω_c. An optimum size of the cascode can be calculated by maximizing the power gain or by minimizing the noise figure. Both will yield different results. The power gain usually benefits from a lower Miller factor since most commonly $(1+M\alpha_{gd})^2 > \left(1+\left(\frac{\omega_0}{\omega_c}\right)^2\right)$ and hence a larger M2 is preferable. When optimizing the size of M2 with respect to noise figure the behavior is somewhat more complex since it affects all noise contributions. Nevertheless the noise figure tends to have a very flat behavior around the optimum. The optimum width of M2 seems again to be larger than M1 (e.g. $W_2 = 2W_1$). However, even doubling or halving the width has a very limited effect of 10 to 20%.

Somewhat unexpectedly, M2 is often chosen smaller than M1. This limits the capacitive load contribution at the output node. Doubling the capacitive load at the output will double the required Q_L of the load inductor to generate the same effective load resistance. Especially when high gain is mandatory and the available Q_L headroom is limited, the output capacitance should be minimized. This capacitive contribution also has a low Q which could lower the effective load resistance and is highly non-linear which could give distortion owing to the large signals present at the output node (cf. Section 4.6). A sound choice for M2 is to make it about half of M1. This limits the capacitive load while still only minorly affecting the noise figure.

4.6 Output Capacitance Non-Linearity

Non-linear components have a more severe influence on the overall linearity as the signal levels are higher. Any non-linearity in the output admittance may play a significant role in the overall IIP3 of the LNA. For the LNA, the main non-linear component in the output admittance is the drain-bulk junction of M2 as indicated in Fig. 4.13. It is assumed for now that this capacitance has no series resistance. This is a worst case assumption for the linearity since any series resistance will reduce the effective capacitive load and improve the linearity. The total output capacitance is denoted C_L and the non-linear fraction ξ_{db2} where index '$db2$' stands for the drain-bulk junction diode of M2:

$$\xi_{db2} = \frac{C_{db2}}{C_L} = \frac{\alpha_{db} C_{gs2}}{C_L} \tag{4.63}$$

The quality factor of the total load network is usually — or at least it should be — determined by the quality factor of the inductor:

$$Q_L = \omega_0 C_L R_L \approx \frac{R_L}{\omega_0 L_d} \approx \frac{\omega_0 L_d}{R_{L,s}}, \tag{4.64}$$

where R_L is the equivalent parallel load resistance and $R_{L,s}$ is the series resistance of the load inductor L_d. The last approximation is valid if $Q_L^2 \gg 1$ and $R_{L|R_{L,s}=0} \gg R_L$. This last requirement states that the equivalent parallel resistance owing to all contributions accept that of the load inductor[4] is much larger than the contribution of the load inductor itself.

[4]These contributions stem from the output resistance of the cascode and the series resistance of C_L as will be explained in Section 4.8.

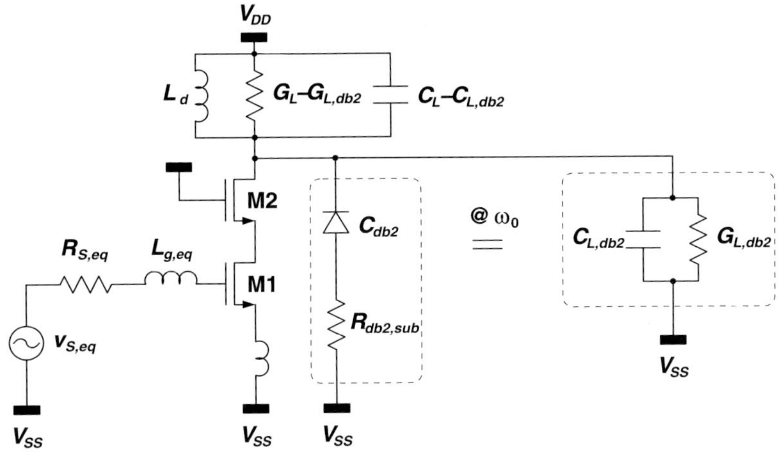

Figure 4.13: *Schematic of the LNA showing the non-linear contribution of C_{db2} to C_L.*

The equation for $\mathrm{IV3}_{db2}$ (voltage amplitude at intercept) for C_{db2} is similar to that of the input diode and given by (4.40) where V_{dO} is replaced by the DC output voltage approximated by V_{DD}:

$$\mathrm{IV3}_{db2} = \sqrt{\frac{4}{3}\left|\frac{(-M_J V_{DD} + V_{bi} + V_{DD})(V_{bi} + V_{DD})^2}{M_J(-M_J^2 V_{DD} + 3M_J V_{bi} + V_{DD} + 3V_{bi})}\right|}. \quad (4.65)$$

However the fundamental output current used for the calculation of (4.65)is the fundamental current through C_{db2} (i.e. $j\omega_0 C_{db2} v_L$). The actual fundamental output current is given by $\frac{2v_L}{R_L}$ since the capacitance is tuned out with L_d. Consequently a correction factor of $\sqrt{\frac{2}{\omega_0 R_L C_{db2}}} = \sqrt{\frac{2}{\xi_{db2} Q_L}}$ needs to be applied to the IV3 value. Referring $\mathrm{IV3}_{db2}$ back to the input of the LNA and converting it to available input power yields

$$\begin{aligned}
\mathrm{IIP3}_{db2} &= 30 + 10\log\left|\frac{4(-M_J V_{DD} + V_{bi} + V_{DD})(V_{bi} + V_{DD})^2}{3M_J(-M_J^2 V_{DD} + 3M_J V_{bi} + V_{DD} + 3V_{bi})}\right| \\
&\quad - 10\log\left(\left(\frac{\omega_T}{\omega_0}\right)^2 \frac{R_L^2}{2R_{s,eq}}\right) - 10\log\left(\frac{\omega_0 R_L C_{db2}}{2}\right) \\
&= 30 + 10\log\left|\frac{4(-M_J V_{DD} + V_{bi} + V_{DD})(V_{bi} + V_{DD})^2}{3M_J(-M_J^2 V_{DD} + 3M_J V_{bi} + V_{DD} + 3V_{bi})}\right| \\
&\quad - 10\log\left(\left(\frac{\omega_T^2}{\omega_0}\right) \frac{R_L^3 C_{db2}}{4R_{s,eq}}\right).
\end{aligned} \quad (4.66)$$

Notice that $\mathrm{IIP3}_{db2}$ is inversely proportional to R_L^3. For larger R_L the output voltage increases linearly. This explains a power of two since $v_L^2 \propto P_{av,s}$. The third power can be explained by

Diode parameter	Value
C_J [F/m^2]	0.0013
V_{bi} [V]	0.88
M_J [ ]	0.43
V_{DD} [V]	1.5

Table 4.3: *Bottom-plate junction parameters of the n+ pwell diode.*

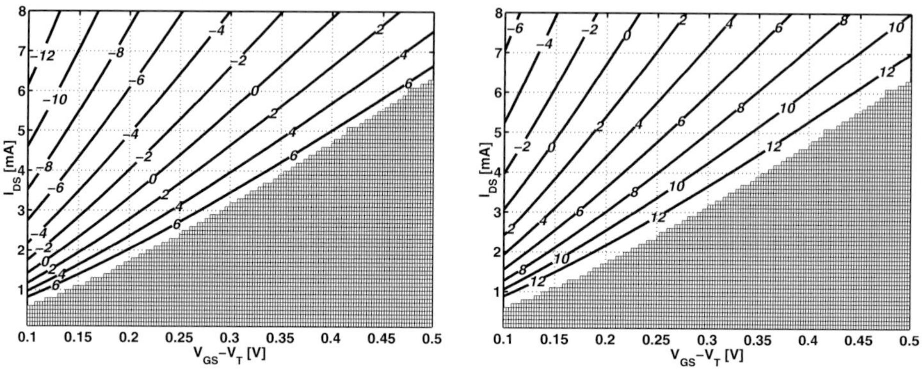

(a) $\tau_{db2} = 0 \left(\frac{\text{rad}}{\text{s}}\right)^{-1}$.

(b) $\tau_{db2} = 2 \times 10^{-10} \left(\frac{\text{rad}}{\text{s}}\right)^{-1}$.

Figure 4.14: *Contour plots of* IIP3_{db2} *for a cascode transistor M2 half the size of M1.*

considering only the output resonant tank. The fundamental component in the output current is inversely proportional to R_L while the third order intermodulation current is not related to R_L. The IIP3 voltage of the output tank itself is therefore inversely proportional to $\sqrt{R_L}$. Consequently, The IIP3 at the input of the LNA is inversely proportional to R_L^3. Numerical calculations in this section were done with a load resistance of 500 Ω. Equation (4.66) was obtained without considering the Miller-effect on M1. This allows comparison with IIP3_{M1} and IIP3_p given by (4.37) and (4.42). Introducing the Miller effect would not yield drastic changes.

Fig. 4.14(a) depicts the IIP3 contribution of C_{db2} for a cascode transistor M2 half the size of M1[5]. The main diode parameters are listed in Table 4.3. Comparison of Fig. 4.14(a) with Fig. 4.10 and Fig. 4.7(f) shows that IIP3_{db2} is actually lowest and hence dominates the LNA IIP3. IIP3_{db2} can be improved by increasing the substrate resistance $R_{db2,sub}$ in series with C_{db2}. This will reduce the effective capacitive load contribution at the output to

$$C_{L,db2} = \frac{C_{db2}}{1 + \omega_0^2 C_{db2}^2 R_{db2,sub}^2} = \frac{C_{db2}}{1 + \omega_0^2 \tau_{db2}^2}. \tag{4.67}$$

The substrate resistance can be increased by placing the substrate contacts sufficiently far from the cascode device or by removing them all together. A realistic value for τ_{db2} is $2 \times 10^{-10} \left(\frac{\text{rad}}{\text{s}}\right)^{-1}$.

[5] Note that doubling the size of M2 results in a IIP3 reduction of 3 dB.

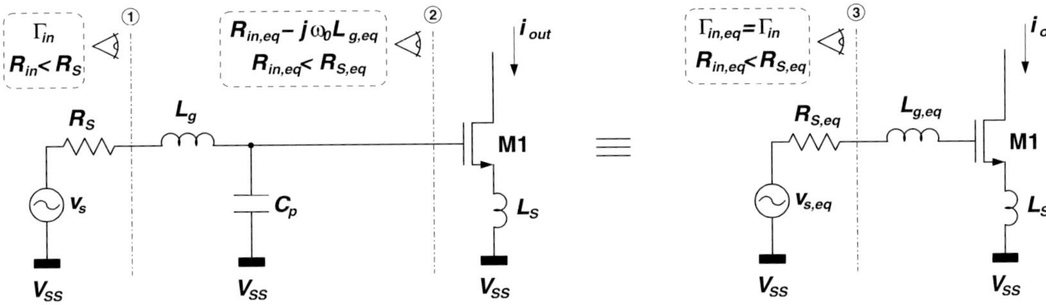

Figure 4.15: *Schematic of the LNA with reduced input resistance $R_{in} < R_S$. It is also indicated that the lossless matching network does not change the reflection coefficient Γ_{in}.*

This amounts to a resistance of 1 kΩ for a drain-bulk capacitance of 100 fF. Substituting C_{db2} with $C_{L,db2}$ in (4.66) yields

$$\begin{aligned} \mathrm{IIP3}_{db2} = 30 + 10\log &\left| \frac{4\left(-M_J V_{DD} + V_{bi} + V_{DD}\right)\left(V_{bi} + V_{DD}\right)^2}{3M_J\left(-M_J^2 V_{DD} + 3M_J V_{bi} + V_{DD} + 3V_{bi}\right)} \right| \\ - 10\log &\left(\left(\frac{\omega_T^2}{\omega_0}\right) \frac{R_L^3 C_{L,db2}}{4R_{s,eq}\left(1 + \omega_0^2 \tau_{db2}^2\right)} \right), \end{aligned} \tag{4.68}$$

and is plotted in Fig. 4.14(b). An improvement of 6 dB is noticed compared to Fig. 4.14(a).

For increasing frequencies the linearity contribution of M1, $\mathrm{IIP3}_{M1}$ was shown to increase quadratically with the frequency. On the other hand, neglecting the change in $R_{S,eq}$, $\mathrm{IIP3}_{db2}$ will improve linearly when $R_{db2,sub}$ is neglected. However, if $R_{db2,sub}$ is taken into account, $\mathrm{IIP3}_{db2}$ will increase with the third power of the frequency (beyond $1/\tau_{db2}$). Indeed, the drain-bulk capacitance becomes progressively more *invisible* with respect to the substrate resistance. Hence, $R_{db2,sub}$ needs to be maximized during the layout stage of the design, especially since the exact value of any substrate resistance is hard to predict. Consequently $\mathrm{IIP3}_{M1}$ will usually become more dominant at higher frequencies. The overall IIP3, expressed in [mW] can be calculated by

$$\frac{1}{\mathrm{IIP3}^2} = \frac{1}{\mathrm{IIP3}_{M1}^2} + \frac{1}{\mathrm{IIP3}_p^2} + \frac{1}{\mathrm{IIP3}_{db2}^2}. \tag{4.69}$$

The actual importance of the three contributions at a specific frequency depends heavily on the design of M1 and M2 and the load resistance R_L. They should all be considered during design and layout and verified with simulations.

4.7 Impact of a Non-Zero S_{11}

Up until now it has always been assumed that the input impedance of the LNA was matched to 50 Ω. This is equivalent to $\Gamma_{in} = S_{11} = 0$ or $|S_{11}| = -\infty$ dB. Obviously in real life this is not achievable. Therefore a margin for matching was introduced which is widely accepted throughout the telecom world:

$$|S_{11}| < -10 \text{ dB}. \tag{4.70}$$

In case the input impedance is real and considering again a 50 Ω source impedance, this can be rewritten as

$$26\ \Omega < R_{in} < 96\ \Omega. \tag{4.71}$$

Now consider again the input of the LNA and let us assume for the moment that $C_p = 0$. If the input impedance is real but not necessarily 50 Ω, then Γ_{in} is real and the power gain of the LNA can be written as

$$G_T = \frac{R_L R_S}{(R_{in} + R_S)^2}\left(\frac{\omega_T}{\omega_0}\right)^2 = (1 - \Gamma_{in})^2\, \frac{R_L}{4R_S}\left(\frac{\omega_T}{\omega_0}\right)^2. \tag{4.72}$$

This means the power gain of the circuit has changed with a factor $(1 - \Gamma_{in})^2$. If $\Gamma_{in} < 0$ or $R_{in} < 50\ \Omega$ this change is actually an increase. For an input resistance of 30 Ω, which implies $\Gamma_{in} = -0.25$ and $20\log(\Gamma_{in}) = -12$ dB, the gain of the LNA is increased with 2 dB.

Hence, By reducing the input impedance of the LNA the power gain can be increased. Even though less power is absorbed by the LNA input, the absorbed power is used more efficiently to generate output current. The presence of a parasitic input capacitance does not change above discussion. Indeed, since L_g and C_p constitute a lossless matching network the reflection coefficient remains unaltered. Equation (4.72) can be reused where R_S is replaced by $R_{S,eq}$. It was further shown in [Jan01] that reducing the input impedance has very little effect on the noise figure. Since both input and output current are larger, all IIP3 contributions referred to the input of the LNA are reduced by the same factor $(1 - \Gamma_{in})^2$. The input quality factor Q_{in} is increased by $(1 - \Gamma_{in})$.

4.8 Output Considerations

4.8.1 Load Impedance Constraints

The power gain of the LNA was shown to be proportional to R_L, the equivalent load resistance. This resistance has several contributions, i.e. different resistors or equivalent resistors connected in parallel to the drain of M2. The main contribution, (that is the lowest resistance) comes from the inductor and is denoted $R_{L,L}$ in Fig. 4.16. It should be the most important one since it is most well known or can be accurately simulated. Indeed, it is calculated as

$$R_{L,L} = R_{L,s}(Q_{L,L}^2 + 1) \approx \frac{\omega_0^2 L_d^2}{R_{L,s}}, \tag{4.73}$$

where L_d is the inductance and $R_{L,s}$ is the series resistance of the inductor. If $Q_{L,L}^2 \gg 1$ then the equivalent parallel inductance can be approximated by L_d. This requirement is usually fulfilled. The value of L_d depends almost solely on the geometry of the coil. $R_{L,s}$ depends both on the geometry and the sheet resistance of the metal layers. The geometry factor can be calculated

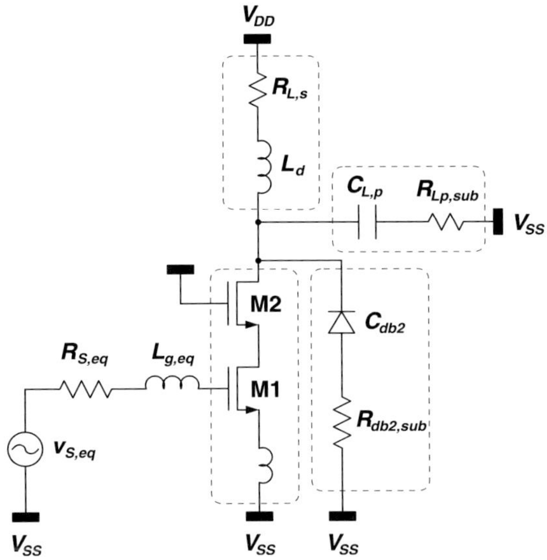

Figure 4.16: *Schematic of the LNA indicating the different contributions to the load resistance.*

arbitrarily accurately by a 2.5D or 3D first-order electromagnetic simulator like FASTHENRY [Kam94]. The metal sheet resistance has a certain standard deviation which is provided by the foundry and can be up to 20% or more.

The coil also causes another contribution which is related to its parasitic capacitance to the substrate, $C_{L,p}$:

$$R_{L,CLp} = R_{Lp,sub} + \frac{1}{\omega_0^2 C_{L,p}^2 R_{Lp,sub}}, \tag{4.74}$$

where $R_{Lp,sub}$ is the series substrate resistance. If the substrate resistance were infinite or zero there would be no extra resistive loading to the output. Unfortunately it is somewhere in between and even more unfortunately often it is close to the worst possible value ($= 1/\omega_0 C_{L,p}$). However, by adding a ground plane beneath the inductor (in metal or poly) the resistance can be reduced to a few Ohm. Consequently $R_{L,CLp}$ will be in the order of tens or hundreds of kilo Ohm and will not significantly affect the load resistance. The capacitive loading however is increased and could limit R_L through self-resonance of the inductor. It should also be noted that adding a full ground shield beneath the inductor would give rise to eddy currents, completely destroying the inductor's Q. This is solved by patterning the ground shield as will be discussed in Section 4.10.2.2.

The third contribution has already been touched upon earlier. This contribution stems from the drain-bulk capacitance of M2. In Section 4.6 it was shown that the substrate resistance of M2 should be maximized to avoid severe linearity degradation. However, by increasing τ_{db2} beyond $\frac{1}{\omega_0}$ the capacitive loading is reduced but the resistive loading is increased:

$$R_{L,db2} = R_{db2,sub} + \frac{1}{\omega_0^2 C_{db2}^2 R_{db2,sub}} \approx R_{db2,sub}, \tag{4.75}$$

Hence it becomes even more important to maximize $R_{db2,sub}$. However it was already stated that $R_{db2,sub}$ is very hard to model or calculate. It can be increased maximally to 10 or 20 kΩ for a substrate resistivity of 20 Ωcm but can be far less.

A final contribution stems from the output impedance of the cascode pair. It is in the order of several tens or even hundreds of kilo Ohm and can usually be neglected. The total load resistance can now be calculated as:

$$\frac{1}{R_L} = \frac{1}{R_{L,L}} + \frac{1}{R_{L,CLp}} + \frac{1}{R_{L,db2}}. \tag{4.76}$$

The total quality factor of the load, assuming $Q^2_{L,L} \gg 1$ is given by

$$Q_L = \frac{R_L}{\omega_0 L_d}. \tag{4.77}$$

Often, in a receiver system design, the gain of the LNA should be within a few dB of a target specification. If it is too high, subsequent stages may give linearity problems. If it is too low they may have noise problems. This means the actual load resistance should be known with an accuracy of a factor two if all other parameters are considered fixed. Consequently R_L should be solely determined by the inductor itself. Especially considering the uncertain contribution of the drain-bulk capacitance of M2, the load resistance will be limited to a few hundred Ohm, maximum 1 kΩ. Even though at higher frequencies, higher quality inductors can be made, the maximum R_L does not increase since $R_{L,Cdb2}$ is already frequency independent and should never become dominant. In practice this load resistance is usually high enough since larger values could compromise the stability due to the large gain.

4.8.2 Output Matching

In essence, no output matching is required from a basic performance standpoint. Often, the LNA directly drives the mixer and the impedance level at the connection can be much higher. However, if the LNA is designed as a stand-alone modular block, it is often more interesting to match the output impedance to 50 Ω. As a consequence the LNA should be a plug'n play stand-in for any other 50 Ω in - 50 Ω out LNA at the same frequency. Moreover this will ease measurements since most RF-measurement tools are equipped with 50 Ω ports. Connecting with 50 Ω cables should theoretically avoid any reflections.

For output matching, R_L needs to be transformed (with minimum losses) to 50 Ω. Several techniques are available. Maybe the most obvious one is adding an output buffer. Clearly this is no real transformation since an active circuit is used. Extra power is consumed and this is not a desirable feature. Consequently, this option will only be used if other techniques fail or have disadvantages that are even worse. Two real on-chip transformers will be discussed here. Although other schemes exist, they are often hard to realize on-chip.

The first circuit is the capacitive divider and is shown in Fig. 4.17(a). It is made up of C_1 and C_2. This topology yields extra capacitive loading $C_{L,div}$ at the drain of M2. The total load capacitance C_L tunes out the load inductance L_d at the operating frequency:

$$\omega_0 = \frac{1}{\sqrt{L_d C_L}}, \tag{4.78}$$

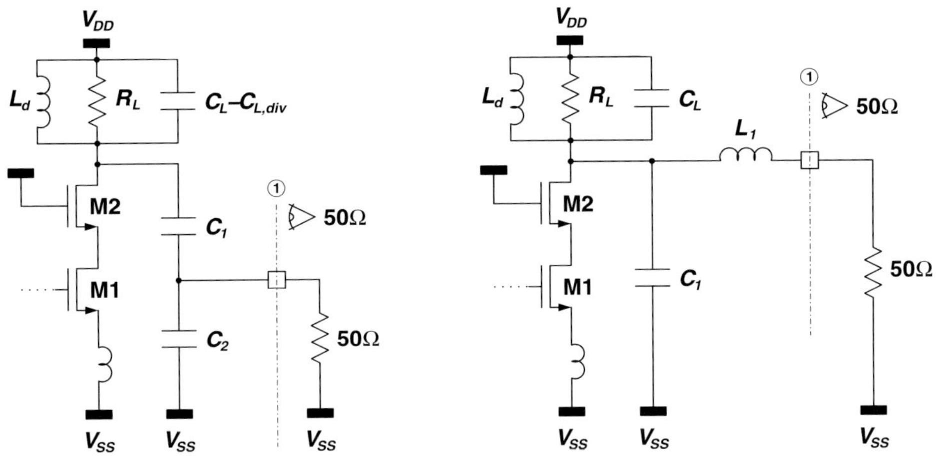

(a) Capacitive divider. (b) L-match.

Figure 4.17: *Two output matching schemes based on impedance transformation.*

where the main contributors to C_L are the parasitic capacitance of the coil and the divider itself. Hence for a given load inductor the amount of capacitance available for the divider is fixed. Or in other words, a specific divider will limit the available inductance. Advantage of this circuit is that the transformation ratio depends mainly on the ratio of C_1 and C_2 if both capacitors are made small. However C_2 also contains the output bonding pad which puts a lower bound on both C_2 and C_1. Hence, depending on the required inductance value this topology may or may not be an interesting choice.

Another matching circuit that is easily implemented is the L-type matching network, and is depicted in Fig. 4.17(b). Unlike the capacitive divider, this topology is not limited by the self-resonance of the load inductor. Indeed, if the circuit is matched, the total capacitance at the drain of M2 is now given by

$$C_{L,tot} = C_L + \frac{1}{\omega_0\sqrt{50R_L}} = \frac{1}{\omega_0^2 L_d} + \frac{1}{\omega_0\sqrt{50R_L}}. \tag{4.79}$$

The inductor L_1 is best implemented as a bonding wire. In that way the losses in the matching circuit are negligible and the bonding pad capacitance can be taken as part of $C_{L,tot}$. If L_1 is realized on chip then extra capacitance is added at both the output and the drain of M2. Consequently the matching network has actually become a Π-type matching network with a lossy inductor which clearly has lesser performance.

4.9 LNA Bandwidth

The bandwidth of the LNA is important for two reasons. It should cover the entire signal band and it should remain to do so considering the spread in process parameters and temperature of operation. One possible requirement could be that the gain in the signal band should not be less than 1 dB of its maximum within the corners of the operation region.

In order to calculate the 3 dB bandwidth of the LNA it is interesting to define the total quality factor of the LNA:

$$Q_{LNA} \triangleq \frac{\omega_0}{3\text{ dB}BW}. \tag{4.80}$$

This quality factor has several different contributions. For a single stage amplifier, it can be found as the sum of an input and output quality factor:

$$Q_{LNA} = Q_{LNA,in} + Q_{LNA,out}. \tag{4.81}$$

This expression is only valid if the input and output networks are centered at the same frequency. The bandwidth can in principle be increased by introducing a frequency offset between input and output resonance. However the resulting power gain will be more sensitive to process variations since they can drive these frequencies further away from each other. This will cause a serious dip in the power gain at what is probably the main frequency of operation, the center of the intended frequency band. In order to decrease this dip, the individual Q-factors should be lowered further which will inevitably decrease the gain of the circuit. That is why both resonant networks are best centered at the same frequency.

Note that $Q_{LNA,in}$ can be different from $Q_{in} = \frac{v_{gs}}{v_s}$ (cf. equation (4.33)). It is defined by

$$Q_{LNA,in} = \frac{\omega_0}{\omega_{gs,-3dB,+} - \omega_{gs,-3dB,-}}, \tag{4.82}$$

where ω_0 is the center frequency at which $Q_{in} = \frac{v_{gs}}{v_s}$ is maximum and $\omega_{gs,-3dB,+}$ and $\omega_{gs,-3dB,-}$ are the upper and lower -3 dB frequencies of Q_{in}. For the L-type input matching network consisting of L_g and C_p, it can be shown that $Q_{LNA,in,L} \approx Q_{in}$. For another type of matching network, this Q factor can be larger as well as smaller.

The output contribution, $Q_{LNA,out}$,

$$Q_{LNA,out} = \frac{\omega_0}{\omega_{out,-3dB,+} - \omega_{out,-3dB,-}}, \tag{4.83}$$

where ω_0 is the center frequency at which the output voltage over the 50 Ω load is maximum[6] and $\omega_{out,-3dB,+}$ and $\omega_{out,-3dB,-}$ are the upper and lower -3 dB frequencies of the output voltage. $Q_{LNA,out}$ depends partly on Q_L, given by (4.77) and will be smaller for a larger inductor with the same R_L. However, $Q_{LNA,out}$ depends also on the actual implementation of the matching network. For the same load network, the capacitive divider from Fig. 4.17(a), generally yields a

[6]The output voltage considered here is calculated for a fixed frequency independent current injected into the load, so independent of the input resonant network.

smaller Q than the L-match from Fig. 4.17(b). The more capacitive headroom that is available for the divider, the smaller $Q_{LNA,out}$ or the larger the corresponding bandwidth.

At the input the availability of different matching networks, especially for on-chip integration is rather limited. Usually, these networks have parasitic losses which can severely deteriorate the performance if not carefully designed. Due to these parasitic losses the input matching network should be as simple as possible. In other words, the less components, the less losses, the better the performance[7]. At the output, the bandwidth of the LNA can be increased by increasing the load inductance for a given load resistance. Increasing the inductance will however lower the self resonance frequency of the inductor and possibly disable matching by means of a capacitive divider. Alternatively, the load resistance itself can be lowered which naturally degrades the power gain of the amplifier. Sometimes even an extra resistor or linear MOS transistor is placed in parallel with the inductor to further lower the Q and reduce the dependence of R_L on the physical properties of the coil. It should also be noted here that it is often advised to lower R_L to avoid stability problems due to the large gain. This also helps to increase the bandwidth.

4.10 Layout Aspects

4.10.1 RF Bonding Pads

A standard bonding pad consists of all metal layers available in the technology which are all connected with numerous vias. In this way any metal can be used to route to the active circuit. Looking at this structure from an RF standpoint reveals a few possible disadvantages. The typical bonding pad is quite large (5000 to 10000 μm^2) and has a non-negligible capacitance towards the substrate. This substrate is connected to ground through the substrate itself which behaves like a resistor. Consequently the bonding pad is a large capacitance (0.1 to 0.5 pF) with a large resistance in series (100 Ω to 5 kΩ).

At the input of the LNA, the bonding pad capacitance increases C_p and the substrate resistance lowers $R_{cp,p}$. Both are unwanted as discussed in Section 4.3.1 and Section 4.3.3. Moreover, any noise in the substrate (especially at high frequencies) can easily couple into the circuit via the bonding pad capacitance. At the output, the extra capacitance will affect the output matching. The substrate resistance will reduce the output power and will make the effective capacitive loading hard to predict. Consequently the capacitance of the bonding pad should be minimized and the corresponding quality factor should be maximized.

By limiting the bonding pad to only the top metal layer, the distance to the substrate is maximized. The size of the bonding pad can be reduced to whatever is allowed by the design rules. In this way the capacitance is reduced. Cutting off the corners of the rectangular pad, which results in an octagonal shape, can further lower the capacitance. Placing a ground shield between the top metal and the substrate further bypasses the large and unpredictable substrate resistance [Rof98b, Col99]. This ground shield is best implemented in the lowest metal layer or in low resistance poly silicon. The first gives the highest Q-factor but the highest capacitance.

[7]Remember that each loss is represented by a resistor and each resistor is accompanied by a proportional noise power source.

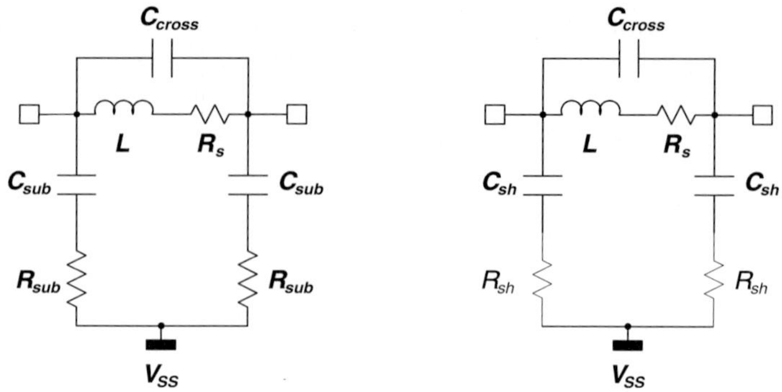

(a) No ground shield. (b) With patterned ground shield.

Figure 4.18: *Models for integrated inductors above a lossy substrate* .

The latter gives a somewhat lower Q-factor with a somewhat lower capacitance. Both are viable solutions.

4.10.2 On-Chip Inductors

4.10.2.1 Modelling

The inductor is by far the most difficult component to realize in standard CMOS technologies. Several different effects and parasitics are of the upmost importance in its design and layout. Both the quality factor and the self resonance frequency are extremely affected by them. A simple but very accurate model for integrated inductors was presented in [Cro96] and is shown in Fig. 4.18(a). It was developed especially for technologies with lowly doped or high resistivity substrates[8].

L is the inductance of the coil. It is mostly dependent on the geometry of the device. R_s is the series resistance of the inductor. It incorporates both the physical resistance in the inductor windings (possibly increased by the skin-effect) and the magnetic losses (eddy currents) in the vicinity of the coil (in the substrate or neighboring metal). C_{cross} models the cross capacitance between the different windings of the inductor. C_{sub} is the capacitance of the coil toward the substrate. It is equally distributed between the two nodes of the device. Since the substrate is high-ohmic, it is connected to ground through a substrate resistance in the order of a few kilo Ohm. This model is fairly accurate up to 3 GHz.

At higher frequencies the accuracy of the model can be increased by repeating the previous model for every section of the inductor and concatenating them. One section can be one tour, half a tour or even just one segment (one eighth of a tour in an octagonal layout). It all depends on the operating frequency and the required accuracy. Note that for every segment separately,

[8]Highly doped substrates are not interesting for integrating inductors since the large eddy currents in the substrate would cause severe losses and completely destroy the quality factor of the inductor.

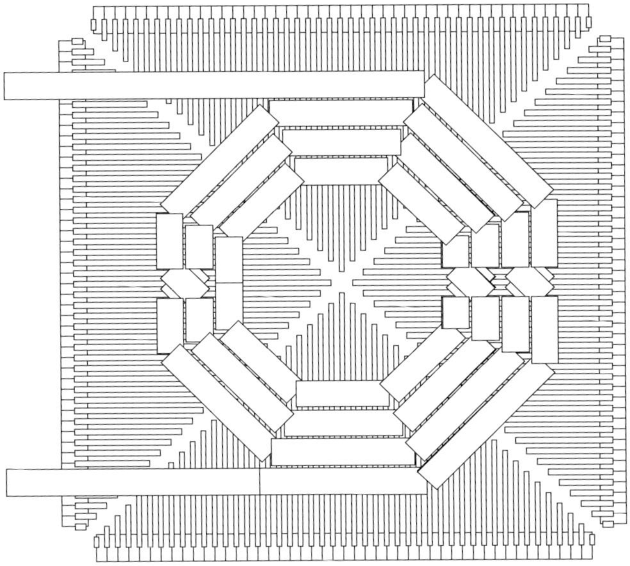

Figure 4.19: *Inductor with patterned ground shield.*

no cross capacitance should be present. They should be added afterwards to the corresponding nodes of any two neighboring segments.

4.10.2.2 Patterned Ground Shields

The self-resonance frequency of the inductor is limited by the capacitors present in the model. The quality factor is determined by the resistive losses in the device modelled by R_s and R_{sub} in Fig. 4.18. R_s is more or less fixed for a given coil geometry since the resistivity and skin-depth of the metal are technology parameters. The magnetic losses due to eddy currents can be minimized by keeping any metal at a sufficiently large distance from the coil. This is why any dummy metal or dummy pwell generation by the mask tools should be avoided. Usually a dedicated layer, explicitly drawn by the designer, will pass this requirement to the mask tools.

The other losses are related to the electric field surrounding the inductor. This electric field is modelled by the capacitors C_{cross} and C_{sub}. It extends into the substrate and causes a current to flow within the substrate. This is modelled by R_{sub}. The severe influence of this resistor was discussed in Section 4.8 and it should be avoided if possible. The electric coupling of the inductor to the substrate can be avoided by placing an electric shield in between and forcing the voltage of the shield to be zero. Analogously to the shield for the bonding pad, this ground shield should also be implemented in the lowest metal layer or in low resistance poly silicon. The first gives the highest Q-factor but the highest capacitance. The latter gives a somewhat lower Q-factor with a lower capacitance. The presence of this shield will alter the model parameters for the coil (Fig. 4.18(b). C_{sub} is increased to C_{sh} since the distance to the shield is smaller than the distance to the substrate. The main advantage is that R_{sub} is now reduced to R_{sh} which is in the order of a few Ohm. The equivalent parallel load resistance contribution is now given by:

$$R_{L,CLp} = R_{sh} + \frac{1}{\omega_0^2 C_{sh}^2 R_{sh}}, \tag{4.84}$$

The question remains whether it is more interesting to use poly or metal as shield. Using poly will somewhat lower the capacitance but increase the resistance. Usually the lower capacitance will not outweigh the higher resistance and the shield is best implemented in metal.

Note that a full ground shield contradicts with the magnetic requirement mentioned earlier, i.e. to keep large metal as far away as possible. Indeed, adding a full ground shield beneath the inductor would cause severe eddy losses; largely reducing the Q of the inductor. This is solved by patterning the ground shield as discussed in [Yue98]. An inductor with a patterned ground shield is depicted in Fig. 4.19. Every rectangle is a segment which is subdivided further for numerical computation by FASTHENRY . The fingers of the ground shield can be realized with the minimum width and minimum spacing allowed by the technology. The tiny currents that can still flow within each finger have a negligible effect on the performance of the inductor. The electric field is still completely intercepted by the ground shield. The electric field lines that fall between two fingers will rather bow towards them than proceed all the way further to the substrate. Consequently the patterning only leads to very small decrease in C_{sh} and a small increase in R_{sh}. The substrate resistance is still completely bypassed.

4.10.3 The Amplifying Transistor

The layout of amplifying transistor M1 is extremely important for the overall LNA performance. Any gate resistance R_{gate} directly adds extra noise to the LNA input:

$$F_{gate} - 1 = \frac{R_{gate}}{R_{S,eq}}. \tag{4.85}$$

The gate resistance can be significantly reduced by using a finger type layout. The gate resistance decreases with the square of the number of fingers since the series connection is replaced by a parallel connection. The distributed nature of the gate resistance leads to an effective gate resistance which is one third of the total gate resistance [Cha91, Raz94, Tin98]. It is given by

$$R_{gate} = \frac{W R_{\square}}{3N^2 L}, \tag{4.86}$$

where N is the number of fingers, $R_{\square}$ is the poly gate sheet resistance, and W and L are the width and length of the complete transistor. Contacting each finger on both sides of the transistor further reduces the resistance by a factor of four since this doubles the effective amount of fingers:

$$R_{gate} = \frac{W R_{\square}}{12N^2 L}. \tag{4.87}$$

In this way the gate resistance can be reduced significantly and the impact on the overall performance is minimized.

Another critical issue to consider in the layout of M1 is the placement of substrate contacts. It is extremely important that the input transistor sees a very "clean" bulk. Any signal on the bulk will result in a feedback current steered from the back gate of M1. This could give rise to increased noise levels, reduced gain and in worst case it could even compromise the stability of the amplifier. One possible precaution is to place an extra grounded guard ring around the transistor. Another technique is to place a n-well guard ring [Cha91]. This will force any substrate current to flow beneath the ring. This will increase the resistance in the path to the bulk of M1 and will urge the substrate currents to flow elsewhere. A combination of these strategies can also be implemented, for instance by placing a grounded ring of substrate contacts both inside and outside the n-well guard ring. In this way, any current that would flow to M1 can be intercepted by the substrate contacts. The effectiveness of these guarding strategies depends a lot on technology parameters like doping levels, n-well depth and the specific geometry of the structure.

4.10.4 The Cascode Transistor

For the cascode transistor, the gate resistance is not so critical since its noise is reduced significantly due to the large impedance at the source of M2. Since the gate voltage of M2 should principally be still, a decoupling capacitor is inserted. The operation frequency determines the size of the capacitor. It becomes smaller for higher frequencies. Even though this capacitor reduces the node impedance at high frequencies, it can dramatically increase the impedance at a lower frequency. Specifically when it goes into parallel resonance with the bonding wire inductor. This extremely high Q resonance may cause instability. The solution is offered by adding some resistance in series with the bonding wire. In this way the quality factor of the resonance can be drastically reduced.

Contrary to M1 where the substrate resistance was minimized, for M2 it is more interesting to maximize the substrate resistance. That way, the non-linear capacitive loading at the drain of M2 can be avoided. This was explained in Section 4.6 and Section 4.8.

4.11 The Common-Gate LNA Revisited

The closer look on the common-source LNA has revealed several important non-idealities, that severely influence the behavior. The rough comparison with the common-gate LNA in Section 2.6 has obviously changed as well. In order to be able to compare both amplifiers, the model of the common-gate LNA introduced in Section 2.6.2 is extended with the same non-idealities as the common-source LNA. This is depicted in the small-signal schematic of Fig. 4.20. However the influence is of much less importance than it was with the common-source LNA. The parasitic input capacitance is tuned out with the source inductor. The NQS gate resistance will appear as an extra resistive load at the input and will usually be much larger than the 50 Ω input impedance.

The new load resistance is found as:

$$R_L = 2\left(ng_m R_{S,nqs} - 1\right) r_{ds}, \tag{4.88}$$

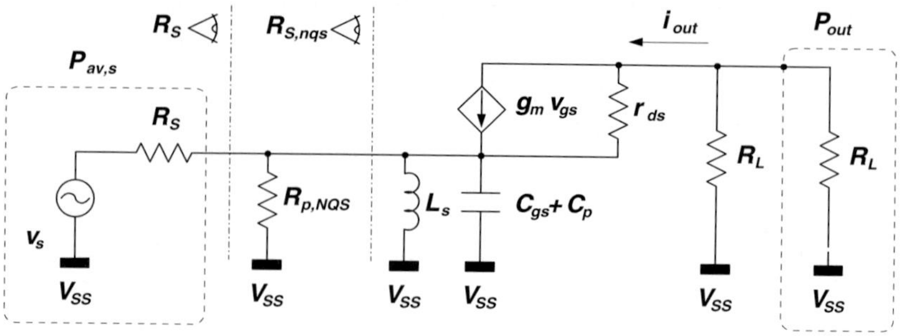

Figure 4.20: *Small-signal schematic of the common-gate LNA including the NQS gate resistance and parasitic input capacitance.*

where $R_{S,nqs}$ is given by

$$R_{S,nqs} = \frac{R_{p,NQS} R_S}{R_{p,NQS} - R_S}, \tag{4.89}$$

and $R_{p,NQS}$ is calculated by (2.53). The gain is slightly reduced since part of the input current is lost through $R_{p,NQS}$:

$$\begin{aligned} G_T &= \frac{R_L}{4R_S} \left(1 - \frac{R_S}{R_{p,NQS}}\right)^2 \\ &= \frac{1}{2} \left(ng_m - \frac{1}{R_S}\right) r_{ds} \left(1 - \frac{R_S}{R_{p,NQS}}\right)^2. \end{aligned} \tag{4.90}$$

An extra term is added to the noise factor, originating from the NQS gate resistance. The total noise factor is approximated by:

$$F \approx 1 + \frac{R_S}{R_{p,NQS}} + \frac{\gamma}{ng_m R_S} + G_T^{-1}. \tag{4.91}$$

Considering only the non-linearity of M1, the IIP3 of the LNA is increased by the same factor that has reduced the gain, i.e.

$$\begin{aligned} \text{IIP3} =& 11.25 + 10 \log \left(\frac{V_{GST} \left(2 + \Theta V_{GST}\right) \left(1 + \Theta V_{GST}\right)^2}{\Theta} \right) \\ & - 20 \log \left(1 - \frac{R_S}{R_{p,NQS}}\right). \end{aligned} \tag{4.92}$$

It is interesting to compare both types of amplifiers for different frequencies in order to be able to choose the best topology for a given application. Naturally, the LNA requirements for a specific application depend not only on the signal levels but also on the architectural choices that were made. Consequently any comparison made here is only illustrative. The designer should

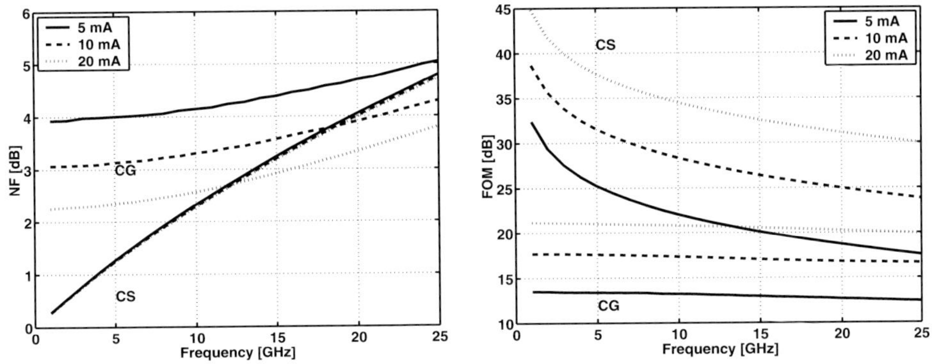

(a) Design for optimum NF. (b) Design for optimum FOM.

Figure 4.21: *Performance comparison of CS and CG LNA as a function of frequency for a 0.25* μm *technology with* $f_{T,max} = 41$ *GHz.*

check whether all required performance parameters can be obtained for the chosen topology. Fig. 4.21 was generated for a 0.25 μm technology. For the common-source LNA, it is assumed that the equivalent source resistance was optimized for noise: $R_{S,eq} = R_{S,opt}$ where $R_{S,opt}$ is calculated by (4.35). This usually implies an extra matching network. Fig. 4.21(a) shows the minimum noise figure attainable at a certain frequency for three different power budgets, corresponding to 5 mA, 10 mA and 20 mA. The noise figure of the common-source LNA is almost independent of the current. This is due to the optimal value of the source resistance which makes the optimum noise figure only depend on $V_{GS} - V_T$ as discussed in Section 4.3.1.2. Obviously at low frequency the common-source amplifier outperforms the common-gate amplifier since the noise factor of the former is proportional to the operating frequency. As an unfortunate consequence, the noise figure of the common-source amplifier increases more rapidly for higher frequencies. The noise figure of both types of amplifiers will cross at a frequency dependent on the current consumption. For a current of 20 mA this crossing is at 10 GHz. For 10 mA it is located at about 16 GHz. It is off-scale for 5 mA. This implies that especially for low frequencies and lower power budgets it is more interesting (at least from a noise perspective) to use a common-source amplifier.

In order to be able to give a somewhat more general performance parameter, the following Figure Of Merit (FOM[9]) is defined:

$$\text{FOM [dB]} = G_T \text{ [dB]} + 10 \log (F - 1) + \text{IIP3 [dBm]}. \tag{4.93}$$

The two topologies have again been optimized within the same power budgets, but this time for maximum FOM. The common-source amplifier scores even better here. Throughout the frequency range it outperforms the common-gate LNA. Again it should be noted that this does not mean that the common-source amplifier is the best choice throughout this frequency range.

[9]The term Figure Of Merit is actually not really appropriate here since the performance is not scaled with the abilities of the technology. Figure Of Performance would be a more fit description.

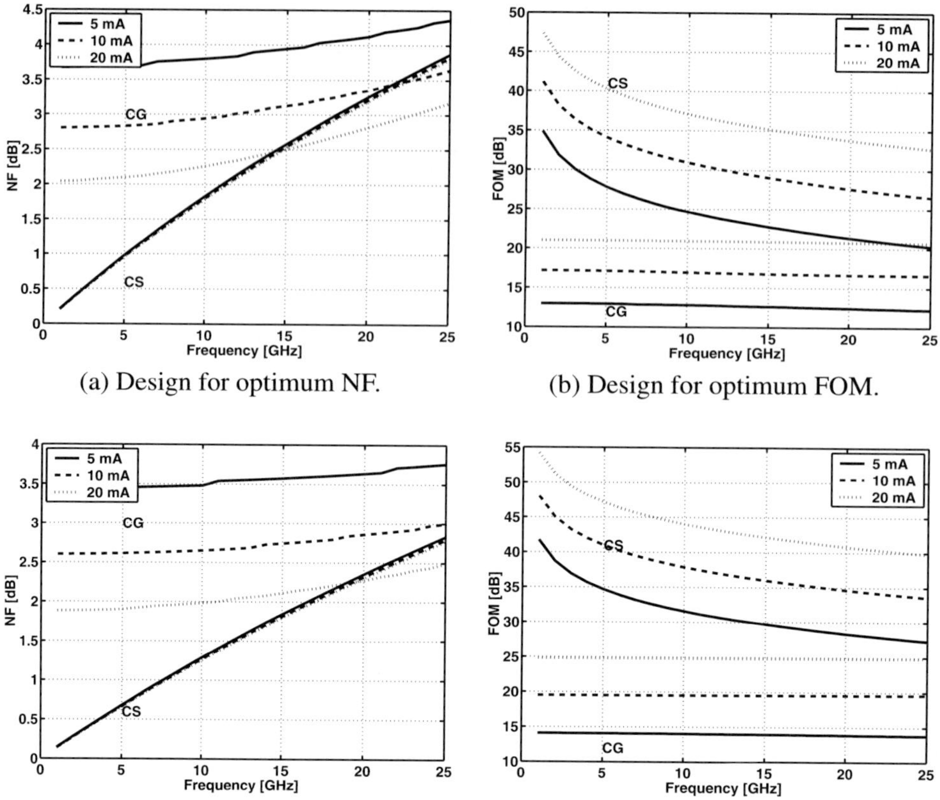

(a) Design for optimum NF.

(b) Design for optimum FOM.

(c) Design for optimum NF.

(d) Design for optimum FOM.

Figure 4.22: *Performance comparison of CS and CG LNA as a function of frequency for a 0.18* μm *technology with* $f_{T,max} = 55$ *GHz ((a) and (b)) and for a 0.13* μm *technology with* $f_{T,max} = 82$ *GHz ((c) and (d)).*

The designer should study the overall receiver performance considering the signal levels of the system and the specifications of the other building blocks. The ultimate goal is to have all possible signals fall within the overall dynamic range of the receiver at a minimum power cost. Also blocking signals need to be taken into account. The optimization strategy is so specific for every application that it really makes no sense to generalize it into one parameter for a given frequency of operation. In this context, the presented FOM is completely arbitrary.

The same plots are repeated for a 0.18 μm technology in Fig. 4.22(a) and (b). The crossing frequency has increased from 10 to 13 GHz for the 20 mA budget and from 16 to 19 GHz for the 10 mA budget. It is again off-scale for 5 mA. This means that the common-source LNA has more to gain from decreasing feature size than its common-gate counterpart. This is due to the inverse proportionality of the noise factor to ω_T (remember: $\omega_T \propto \frac{1}{L^x}$ with $1 < x < 2$). This trend is continued for the 0.13 μm technology plotted in Fig. 4.22(c). The crossing frequencies are now 16 and 22 GHz respectively. The behavior of the minimum FOM for these technologies are indicated in Fig. 4.22(b) and (d). The common-source amplifier has a higher FOM for both technologies.

4.12 Conclusion

This chapter was devoted to providing a rigorous analysis of the common-source LNA with inductive degeneration. All relevant parameters and effects, both inherent and parasitic have been introduced gradually. Their influence on the operation and performance has been evaluated with respect to input matching, noise figure, gain and linearity. Each time conclusions were drawn as to the impact of the respective parasitic on the design of the amplifier. They have been visualized with performance contours in the design space of the amplifying transistor for an illustrative design at 1.5 GHz.

The first parasitic, introduced in Section 4.2, is the non-quasi static effect of the gate-source capacitance of M1. This effect has been modelled with a resistance ($r_{g,NQS}$) and a noise voltage source in series with C_{gs}. The main result of this effect is an extra term added to the noise factor of the amplifier.

In Section 4.3 the impact of the parasitic input capacitance has been investigated. First this capacitance was considered linear and with an infinite Q-factor. It has been incorporated in the LNA model by changing the equivalent source resistance seen by the LNA. The higher C_p, the higher $R_{S,eq}$. This has revealed that C_p has a serious influence on the overall performance. The influence of C_p on the noise figure has been clarified by looking at the relative position of $R_{S,eq}$ with respect to the optimal source resistance $R_{S,opt}$. Usually $R_{S,eq} > R_{S,opt}$ and additional capacitance will increase the noise figure. Moreover it was shown that C_p reduces the gain by lowering the input signal efficiency with which the output current is generated. Since, the signal loss occurs ahead of the non-linear v_{gs}-i_{ds} conversion, the IIP3 of the LNA is improved with the same factor. The OIP3 remains unchanged.

The finite quality factor of C_p was shown to be negligible at moderate frequencies but could become important at 5 GHz and beyond, depending on the different contributions to the capacitance and their respective Q-factor. The non-linearity of C_p has been evaluated as well and is usually negligible, even at higher frequencies.

The effect of the gate-drain capacitance has been evaluated in Section 4.4. C_{gd} has been split up in its Y parameter equivalent yielding four different components: an input and output loading capacitance, a feedback current and a feed-forward current. The input loading capacitance is easily incorporated in C_p, the output loading just adds to the cascode node capacitance and the feed-forward zero is neglected. The influence of the feedback current can be incorporated in the behavioral model by replacing C_{gs} by $C_{gs}(1 + M\alpha_{gd})$. The result of the feedback is equivalent to a reduction in the unity gain frequency f_T.

Section 4.5 has discussed the optimization of the cascode transistor. The size of this transistor has a non-negligible influence on the gain and noise figure of the circuit. Both optimizations would yield different results. Often the size of M2 is chosen significantly smaller than both optimum values in order to reduce the output loading capacitance of the drain-bulk junction of

M2. The influence on the noise figure and gain is minor.

The non-linearity of the output capacitance is analyzed in Section 4.6. This non-linear part stems from the drain-bulk capacitance of M2. It has been shown that this contribution to the non-linearity of the LNA is very important and can even be dominant. The effective capacitive loading can however be reduced by increasing the substrate resistance in series with this capacitance. The IIP3 power is inversely proportional to the effective drain-bulk capacitance. At higher frequency, this capacitance becomes increasingly more invisible and this non-linearity contribution becomes less important.

In Section 4.7 the impact of a finite input match on the performance is evaluated. It is concluded that a finite S_{11} of -12 dB can produce a gain increase of 2 dB without significantly altering the noise behavior. Several output impedance considerations are discussed in Section 4.8. The constraints on the equivalent load resistance have been evaluated. Two different output matching networks have been discussed. In Section 4.9, the different bandwidth limitations have been linked to the bandwidth requirements of a typical application, also taking into account the yield requirement given process variations. Section 4.10 has demonstrated the importance of the layout of the different devices: transistors, inductors, capacitors, even bonding pads.

This chapter concluded with a renewed comparison of the CS and CG LNA when the same or similar parasitics are taken into consideration. The comparison has revealed that the CS LNA has superior performance at low frequency. However the performance degrades more rapidly at increasing frequency than that of the CG LNA. Consequently, depending on the current budget, a cross-over frequency can be identified beyond which the CG LNA performs better. This frequency is located higher for lower power consumption and smaller technologies. Hence, the CG amplifier is most relevant when the allowed power consumption is high and the financial budget is low.

Chapter 5

RF-ESD Co-Design for CMOS LNA's

5.1 Introduction

Even though every pin on a chip is intended to connect to the outside world, they all have a different affinity for ESD-stress. Some pins are more intrinsically immune since they connect for instance to a large junction diode or they feature a large bias resistance in series. The most sensitive pins are the ones connecting directly to the gate of a MOS transistor. Since the LNA input pin connects to the gate of the amplifying NMOS transistor, it is extremely sensitive to ESD. Hence when talking about RF-ESD co-design for LNA's, the main issue is how to protect the input gate without severely deteriorating the performance of the LNA. Although this is a critical issue, very few LNA's have been published with ESD-protection results.

In fact, one of the main bottlenecks for introducing CMOS RF circuits to the market is their susceptibility to ESD. It is mainly due to both gate oxide breakdown and junction degradation related problems. These problems become even more severe as technologies scale further towards nanometer dimensions. As gate length decreases, so does the oxide thickness reducing the breakdown voltage of the transistor gate. The breakdown voltage for a given CMOS technology can be approximated by either of the following expressions:

$$V_{bd} \approx t_{ox} \times 1V/nm \approx L_{min} \times 20V/\mu\text{m} \approx 2V_{max}. \tag{5.1}$$

This yields a gate oxide breakdown voltage of about $5\ V$ for a typical $0.25\ \mu m$ technology. Smaller technologies also feature increased doping levels in order to avoid problems like punch-through and latchup. These decreased doping levels give rise to smaller junction depths (which is actually the primary intent). This in turn reduces the breakdown voltage of the junctions [Ame99].

The ESD problems are still aggravated by the tight design window for the high performance RF circuits, not allowing large ESD devices to be used as protection elements [Rad01]. Most CMOS ESD-protection structures (e.g. as they are used in digital CMOS) have parasitics that are detrimental for the LNA performance (cf. Section 3.3.2.1). They commonly feature two large clamping devices with a current limiting resistor in between. The resistance added at the input (up to a few hundred ohms), would be detrimental for the noise figure of the LNA. The introduced parasitic input capacitance also has a serious influence on the performance of the common source LNA. This influence was explained in Section 4.3. The discussion on RF-ESD co-design will concentrate mainly on this topology.

In order to clarify the ESD-protection methodology described in this chapter, it is interesting to draw an analogy with a totally different branch of science: namely chemistry. In Dutch this science is also called *'scheikunde'* or literally *'the art of separation'*. One of the main statements there is that in order to be able to separate two types of matter they need to have an identifiable difference. This difference can then be used to perform the separation. In chemistry, this difference can relate to density, boiling or melting temperature, solubility or any other physical or chemical property.

Now let us return to the problem at hand. The input of the LNA is facing two types of signals. The RF-signal which should be maximally absorbed by the amplifying device. And the hazardous ESD-signal which should be kept away from the amplifying device. So in fact a separation of these two signals needs to be performed. Consequently one should inspect the differences between them. The first characteristic which is most often used to do the separation is the level of energy in both signals. Indeed the ESD energy is orders of magnitude larger. The separation is then done by using a certain trigger mechanism. Most often this trigger is a voltage level. Once the voltage passes the trigger level, the ESD-protection device is activated and sinks the large-energy current. This trigger device can be for instance, a diode, a bipolar transistor, or a thyristor. All these devices have the characteristic of turning on quickly, once the voltage passes the trigger or threshold level. Disadvantage of this technique is that still large devices are required which have large parasitics compromising the RF-performance. Using these parasitics in the optimization of the LNA results in several co-design methodologies which will be discussed in Section 5.2 and Section 5.3.

Another characteristic difference between the two signals is their frequency content. The ESD-signals have a relatively low frequency up to a few tens of MHz. The RF-frequency is in the GHz range and is continuously increasing due to the quest for larger bandwidth, and enabled by the technological evolution. Consequently, it should be possible to use a simple passive filter-splitter to do the separation. Moreover as the RF-frequency increases the difference will become more pronounced and this technique will become more attractive. A detailed implementation is discussed in Section 5.4.

5.2 ESD-protection within an L-Type Matching Network

5.2.1 Introduction

The classical ESD-protection design is focussed at minimizing the performance degradation induced by the extra ESD-devices. The decay in performance is mainly due to the additional parasitic input capacitance. Therefore, one of the main aims of the classical strategy is to limit the amount of capacitance for a given protection requirement, i.e. for a given ESD-current that can be handled. Even though this strategy is useful and applicable for different types of LNA's, it will be discussed here for the CS LNA with inductive degeneration.

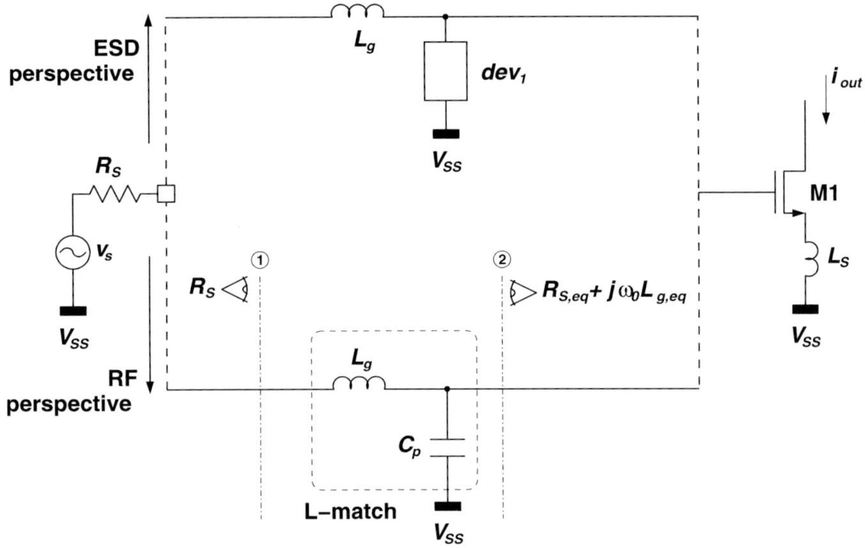

Figure 5.1: *Input ESD-protection within an L-type matching network, both from an ESD- and RF-perspective.*

5.2.2 General Performance

The influence of parasitic input capacitance was discussed thoroughly in Section 4.3. It was explained that the gate inductor and the parasitic input capacitance constitute a lossless L-type matching network as indicated in Fig. 5.1. The resulting performance has been described by means of the equivalent source resistance seen by the gate of M1. This parasitic input capacitance incorporates the input loading of the gate-drain capacitance, the bonding pad capacitance, the wiring capacitance and the parasitic capacitance of the ESD-protection. Thus, the influence of any ESD-protection network can be described as in Section 4.3. This includes the finite Q of the ESD-device, and its non-linearity; both of which are often negligible. It was shown that increasing C_p lowers the gain of the circuit, and increases the noise figure (if $R_{S,eq} > R_{S,opt}$ which is usually the case). Hence it is of the upmost importance that all contributions to C_p are minimized in order to create a sufficiently large headroom for the capacitance of the ESD devices. The gate-drain capacitance of M1 is fixed for a given device. The wiring can be minimized by smart layout. The pad capacitance is minimized by using only the top metal layer and by employing an octagonal layout as discussed in Section 4.10.1.

As an example, the noise figure and power gain of a 1.57 GHz LNA are plotted vs. the capacitance of the ESD-protection device in Fig. 5.2(a). The other contributors to the capacitance ($\sim$ 150 fF) are also taken into account in these simulations but they are not included in the value of C_{ESD}. The current consumption of the LNA was fixed to 6 mA. It is seen in this figure that the NF increases from 0.9 dB to 1.5 dB for an ESD-capacitance of 350 fF. The power gain decreases from 19.3 dB to 15.7 dB. Beyond this capacitance value the circuit can no longer be matched

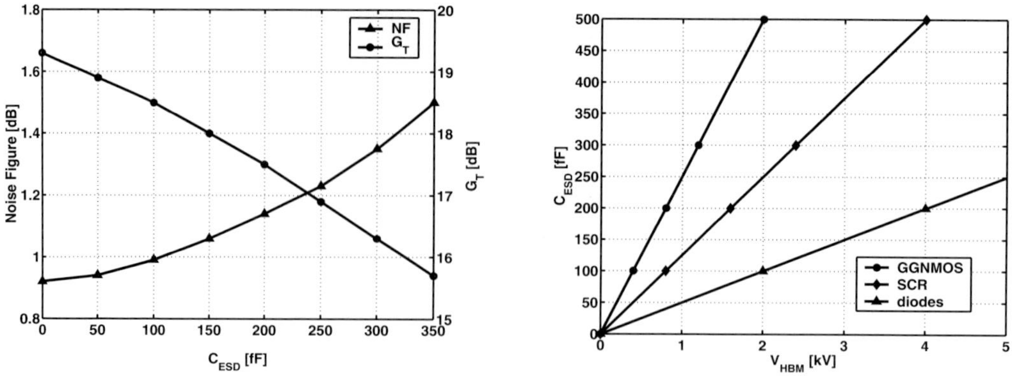

(a) Influence of C_{ESD} on G_T and NF. (b) Required C_{ESD} for different devices.

Figure 5.2: *Impact of the ESD-protection requirements on the gain and noise figure for various ESD devices.*

	Without ESD-protection	**With 100 fF ESD-protection**
Power	9 mW	9 mW
Current	6 mA	6 mA
Supply voltage	1.5 V	1.5 V
S_{11}	19.3 dB	-14 dB
S_{21}	19.3 dB	18.4 dB
S_{12}	<-30 dB	<-30 dB
S_{22}	-23 dB	-25 dB
NF	0.9 dB	1.0 dB
IIP3	-6.8 dBm	-5.1 dBm

Table 5.1: *Comparison between the simulated RF-performance of the LNA with and without input ESD-protection.*

to 50 Ω without an external matching network. For higher frequencies this effect becomes more pronounced, i.e. the curves will be steeper and the cut-off capacitance becomes lower. These curves can be redrawn for different power budgets. Table 5.1 shows a comparison between the main simulation results with and without ESD-protection. This table confirms the theoretical discussion given above. Adding the ESD-protection at the input reduces the power gain from 19.3 dB to 18.4 dB. The noise figure increases from 0.9 dB to 1 dB. The IIP3 improves from 6.8 dBm to 5.1 dBm.

Above discussion shows that it is imperative that the ESD-capacitance is limited. Consequently we need to find the best (= lowest capacitance per current) ESD-device for the job. Fig. 5.2(b) shows a typical plot of the ESD-capacitance as a function of HBM protection voltage for different devices. Together with the contour plots from Section 4.3, this plot allows the designer to choose the amount of ESD-protection required and tolerable for a specific application.

Performance parameter	Design equation
$F\,[\,]$	$1+\frac{\gamma(1+M\alpha_{gd})^2}{\alpha}\left(\frac{\omega_0}{\omega_T}\right)^2\left(g_m R_{S,eq}+\frac{2}{\kappa}\right)+\frac{\alpha\delta(1-\lvert c\rvert^2)}{\kappa g_m R_{S,eq}}+G_T^{-1}$
$G_T\,[\,]$	$\frac{R_L}{4R_{S,eq}}\left(\frac{\omega_T}{\omega_0}\right)^2\frac{1}{(1+M\alpha_{gd})^2}$
$Q_{in}\,[\,]$	$\frac{1}{2\omega_0 C_{gs}(1+M\alpha_{gd})\sqrt{R_{S,eq}R_S}}$
IIP3 [dBm]	$5.25+10\log\left(\frac{V_{GST}(2+\Theta V_{GST})(1+\Theta V_{GST})^2}{\Theta}\right)-20\log(Q_{in})$

Table 5.2: *Main design equations for the common source LNA with an L-type input match.*

The diodes seem to offer the best performance.

The finite quality factor of the ESD-protection device can usually be neglected since both ESD- and RF performance strive for a low resistance. A typical value of a few Ohm for 100 fF is indeed negligible. The main design equations for the L-type CS LNA are repeated in Table 5.2. In the next section, the design, non-linearity and layout of the ESD-diodes are studied.

5.2.3 Design and Layout of the ESD Protection Diodes

The input ESD protection network is shown in Fig. 5.3. It consists of two diodes, D1 and D2, between the RF input and the power supply busses. The use of diodes was based on the fact that they are very efficient and robust ESD devices. Furthermore, their characteristics are fairly simple to model and simulate, allowing a reliable sizing of these devices.

Key considerations in the design of input diodes D1 and D2 is their capacitance, and their high frequency and high current resistance. The first two affect the RF performance of the LNA (as shown earlier), the latter influences the ESD performance of the circuit and, in particular, the bias developed on the input node/gate of the LNA during the ESD pulse. Both the capacitance and the resistance should be minimal, requiring suitable diode optimization. The diodes can be laid out as one finger or multiple finger devices. The junction capacitance of the diode is related mostly to the bottom plate of the diode. The current however, flows mostly through the sidewall. Hence to minimize the capacitance and the resistance of the diode we need to minimize the area and maximize the perimeter. This clearly pleads in favor of a multiple finger structure. Moreover, a one finger diode may give current crowding at the corners, yielding dangerous hot spots where breakdown may initiate. To avoid this the diodes are best implemented with several fingers. These fingers may be realized either as squares or stripes. Stripes give the advantage of a very low resistance since the average distance from a p+ to n+ contact is minimal while still providing sufficient contacts per finger. On the other hand, current crowding and the resulting hot spots may occur at the small ends of the stripes. This problem may be somewhat alleviated by removing the contacts closest to the end of the stripe in order to increase the resistance in

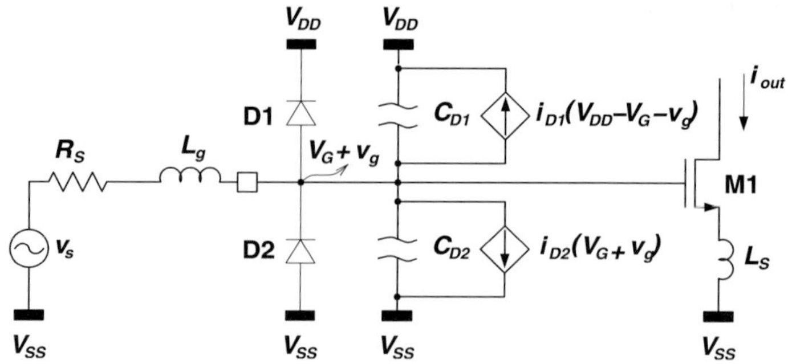

Figure 5.3: *Schematic of the LNA input indicating the non-linearity of the input ESD-diodes*

that path preventing current crowding. When using squares, the average resistance will be a little higher than with stripes but a uniform current distribution is more easily obtained.

5.2.4 Non-Linearity of Input ESD Protection Diodes

The non-linearity of the parasitic input capacitance has been discussed in Section 4.3.2. It was assumed that part of the input capacitance stems from the junction capacitance of a reverse diode. It has been shown that this non-linearity can usually be neglected. In practice, the parasitic diode capacitance is not coming from one but two reverse diodes in opposite configuration as depicted in Fig. 5.3. This feature presents a possible degradation to the linearity performance. Suppose the voltages over the diodes D1 and D2 were chosen identical ($V_G = V_{GS} = V_{DD} - V_G$ or $V_G = V_{DD}/2$). Writing the capacitance of the diodes as a Taylor expansion around the DC gate voltage V_G yields

$$C_{D1} = A_D C_J \left(1 + \frac{\frac{V_{DD}}{2} - v_g}{V_{bi}}\right)^{-M_J} = \sum_{i=0}^{\infty} c_i \left(-v_g\right)^i , \tag{5.2}$$

$$C_{D2} = A_D C_J \left(1 + \frac{\frac{V_{DD}}{2} + v_g}{V_{bi}}\right)^{-M_J} = \sum_{i=0}^{\infty} c_i \left(v_g\right)^i . \tag{5.3}$$

The total diode capacitance C_D is then given by

$$C_D(v_g) = C_{D1} + C_{D2} = 2\sum_{i=0}^{\infty} c_{2i} \left(v_g\right)^{2i} . \tag{5.4}$$

The capacitance is an even function of the small signal gate voltage v_g as shown in Fig. 5.4(a). The capacitance is symmetrical around $V_{DD}/2$. The small signal current through the diodes i_D is found as

$$i_D = i_{D1} + i_{D2} = sC_D v_g = 2s\sum_{i=0}^{\infty} c_{2i} \left(v_g\right)^{2i+1} . \tag{5.5}$$

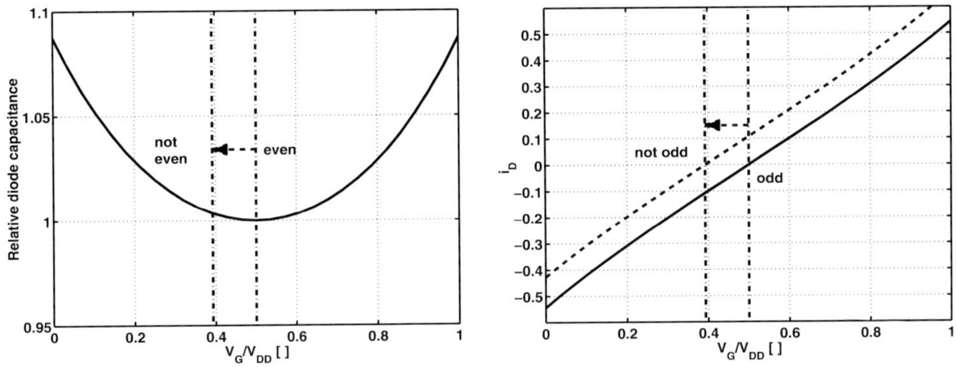

(a) Relative total capacitance. (b) Small signal diode current.

Figure 5.4: *Total diode capacitance linearity for* $V_{DD} = 1.5\ V$.

This implies that the diode current is now an odd function of the gate voltage (around $V_{DD}/2$). No even order distortion is present. The current is shown in Fig. 5.4(b). The coefficients of first and third order are identical to the case of a single reverse diode (cf. Section 4.3.2) since the symmetrical placement of the diodes only affects the even distortion. Consequently, the third order intermodulation voltage IV3 [V amp] of the two diodes is still given by equation (4.40):

$$\mathrm{IV3}_{p,g} = \sqrt{\frac{4}{3}\left|\frac{c_0}{c_2}\right|} = \sqrt{\frac{4}{3}\left|\frac{(-M_J V_G + V_{bi} + V_G)(V_{bi} + V_G)^2}{M_J(-M_J^2 V_G + 3M_J V_{bi} + V_G + 3V_{bi})}\right|}. \tag{5.6}$$

This means the IV3 depends only on the ratio of the zeroth and second order coefficients of the total diode capacitance. The corresponding input referred $\mathrm{IIP3}_p$ can be found by applying the straightforward calculations in Section 4.3.2.

In the most common case where V_G is different from $V_{DD}/2$, the IV3 will however be different. Equations (5.2) and (5.3) can be rewritten as

$$C_{D1} = A_D C_J \left(1 + \frac{V_{DD} - V_G - v_g}{V_{bi}}\right)^{-M_J} = \sum_{i=0}^{\infty} c_i \left(V_{DD} - 2V_G - v_g\right)^i \tag{5.7}$$

$$C_{D2} = A_D C_J \left(1 + \frac{V_G + v_g}{V_{bi}}\right)^{-M_J} = \sum_{i=0}^{\infty} c_i \left(v_g\right)^i. \tag{5.8}$$

For $V_G = V_{DD}/2$ they reduce again to (5.2) and (5.3). Adding both capacitances now yields a total capacitance $C_D = C_{D1} + C_{D2}$ given by:

$$\begin{aligned} C_D(v_g) &= A_D C_J \left(\left(1 + \frac{V_{DD} - V_G - v_g}{V_{bi}}\right)^{-M_J} + \left(1 + \frac{V_G + v_g}{V_{bi}}\right)^{-M_J}\right) \\ &= 2\sum_{i=0}^{\infty} c_{2i} \left(v_g\right)^{2i} + \sum_{k=0}^{\infty} (-v_g)^k \left(\sum_{i=k+1}^{\infty} \binom{i}{k} c_i \left(V_{DD} - 2V_G\right)^{i-k}\right). \end{aligned} \tag{5.9}$$

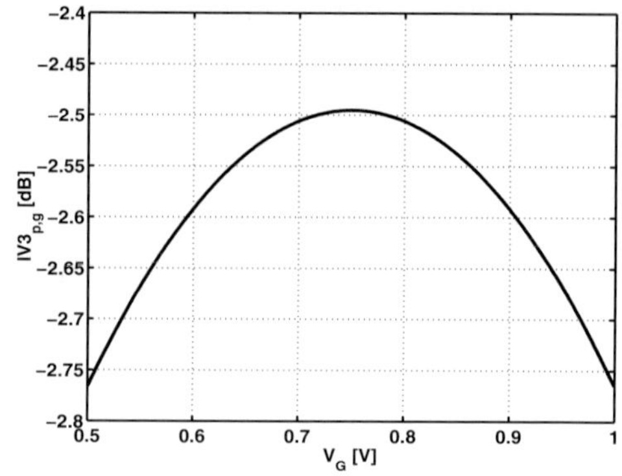

Figure 5.5: $IV3_{p,g}$ *as a function of the DC gate voltage* V_G *for a supply of 1.5 V.*

The capacitance is no longer even and consequently the diode current is no longer odd:

$$i_D = 2sC_Dv_g = 2s\sum_{i=0}^{\infty} c_{2i}\left(v_g\right)^{2i+1} + sv_g\sum_{k=0}^{\infty}\left(-v_g\right)^k\left(\sum_{i=k+1}^{\infty}\binom{i}{k}c_i\left(V_{DD}-2V_G\right)^{i-k}\right). \quad (5.10)$$

Compared to (5.5), an extra third order distortion term is present generated by 4th and higher order terms in (5.7). The third order intermodulation voltage IV3 [V amp] of the two diodes can now be written as

$$\mathrm{IV3}_{p,g} = \sqrt{\frac{4}{3}\left|\frac{2c_0+\sum_{i=1}^{\infty}c_i\left(V_{DD}-2V_G\right)^i}{2c_2+\sum_{i=3}^{\infty}\binom{i}{2}c_i\left(V_{DD}-2V_G\right)^{i-2}}\right|}. \quad (5.11)$$

A closed analytical solution for IV3 is calculated in a straightforward way by using the second derivative of the first line in (5.9) and evaluating it in $v_g = 0$. The IV3 can then be found by using (2.36) where

$$a_1 = C_D(0) \quad (5.12)$$

$$a_3 = \left.\frac{\partial^2 C_D}{\partial v_g^2}\right|_{v_g=0} \quad (5.13)$$

and is completely equivalent to (5.11). The resulting expression is however rather large and yields no further insight. A typical behavioral plot of the IV3 versus the gate voltage is shown in Fig. 5.5. The IV3 is maximum and thus optimal at a gate voltage of 0.75 V which is half of the supply voltage. There, the IV3 is completely identical to the case of a single reverse diode. Any offset from half of the supply yields a linearity degradation. However, even with a gate voltage offset of 0.25 V the IV3 only changes with a few tenths of a dB. We can conclude that the discussion from Section 4.3.2 about the relative importance of the non-linearity of the input capacitance remains valid for two opposite reverse diodes. Under normal circumstances it can be neglected.

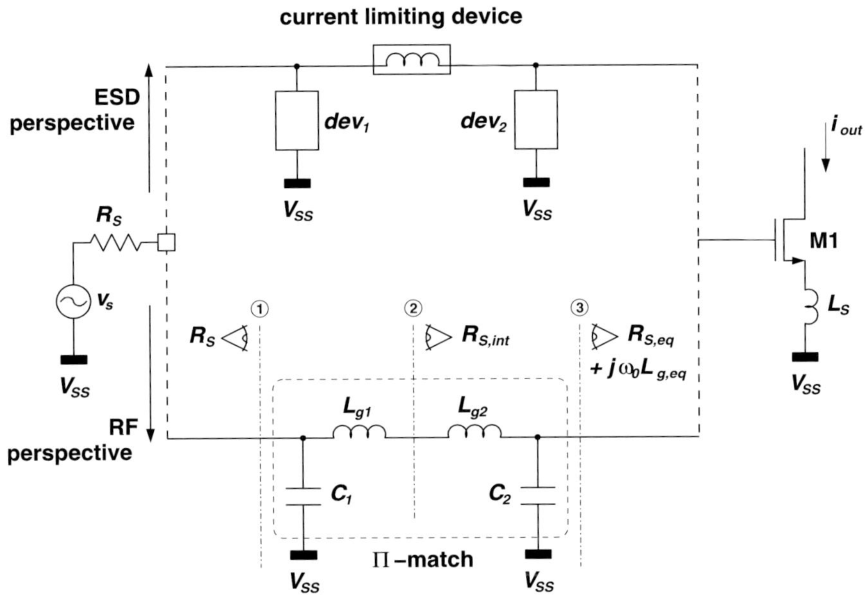

Figure 5.6: *Input ESD-protection within a Π-match, both from an ESD- and RF-perspective.*

5.2.5 Conclusion

The most interesting network for ESD-protection at RF when no topological changes are made to the circuit consists of p+ n-well diodes. The input protection can be realized with one reverse biased diode to ground and one to V_{DD}. A finishing supply clamp yields a fully bidirectional protection with a minimum of capacitive loading. Drawback of this strategy is that at increasing frequency the amount of capacitance that can be tolerated drastically decreases. It can be used up to frequencies of a few GHz depending on the allowed power consumption. In the following sections, different RF-ESD co-design concepts will be introduced and discussed which do not have these high frequencies limitations.

5.3 ESD-Protection within a Π-Type Matching Network

A different ESD-protection network, that can be very appropriate for the input of the LNA is depicted in Fig. 5.6. Instead of the single ESD-device in the L-type network, there are now two clamping devices with some current limiting device in between them. This schematic shows a remarkable resemblance to the standard I/O-protection scheme shown in Fig. 3.15. There however the current limiting device was implemented by a large resistance. For low-noise amplifiers, this resistance cannot be tolerated. It is replaced by an inductor. This inductor has only a small impedance at ESD frequencies and will not be a very efficient current limiter. Still, any impedance is welcome.

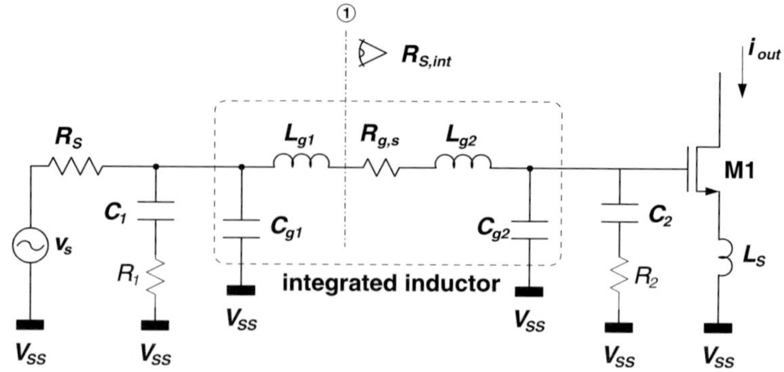

Figure 5.7: *The* Π*-type matching network at the LNA input with the main parasitic components.*

Now let us look at the schematic from an RF standpoint as indicated in the bottom equivalent of Fig. 5.6. The three devices are simplified to their main parasitic at the RF frequency. Both clamping devices are replaced by capacitors C_1 and C_2. The inductor is split into two inductors, L_{g1} and L_{g2}. Since these components are now purely reactive, they constitute a lossless CLC matching network, a Π-match. The influence of this matching network can be incorporated in the LNA model by investigating the equivalent source impedance seen by the gate of M1. The equations for noise figure gain and linearity, derived in Chapter 4 can be reused. However, since an extra component is added at the input, an extra degree of freedom is created for the input match. The different contributions to the quality factor of the LNA were described in Section 4.9. For the Π-type matching network, $Q_{LNA,in}$ differs from its L-match counterpart. For the L-type match, $Q_{LNA,in}$ was approximately equal to Q_{in}. This is no longer the case, since $Q_{LNA,in}$ also depends on $R_{S,int}$, the 'internal' source resistance seen though reference plane ② in Fig. 5.6.

The equivalent source resistance presented to the LNA is now decoupled from the bandwidth associated with the input match. The bandwidth can be increased by increasing $R_{S,int}$. However since $R_{S,int}$ is by construction always smaller than R_S, the bandwidth of the Π-match is always smaller than that of the L-match. The main advantage of the Π-match is that for any value of ESD-capacitance (i.e. any size of the ESD-devices), the equivalent source resistance $R_{S,eq}$ can still be set to any value fitting the required performance. It is just a matter of distributing the ESD devices correctly over C_1 and C_2. In this way it remains interesting from an RF standpoint to reduce the total capacitance (C_1+C_2) but only because a higher capacitance reduces the bandwidth of the circuit. Gain, noise figure and IIP3 can be set independent of the total capacitance.

Up till now the Π-match has been considered ideal and lossless. So what happens if the non-idealities are taken into account. The main non-idealities relate to the on-chip inductor L_g. This is illustrated graphically in Fig. 5.7. Even for an ideal Π-match, it is interesting to make C_1 as small as possible. This will increase the intermediate equivalent resistance $R_{S,int}$ and widen the bandwidth of the circuit. Considering the non-ideal integrated inductor an even more important motivator surfaces. The noise introduced by the series resistance of the inductor adds

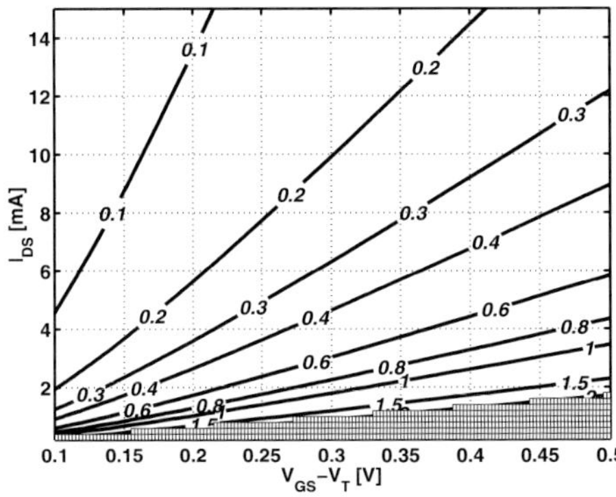

(a) Noise due to gate inductor: $F_{L_g} - 1$.

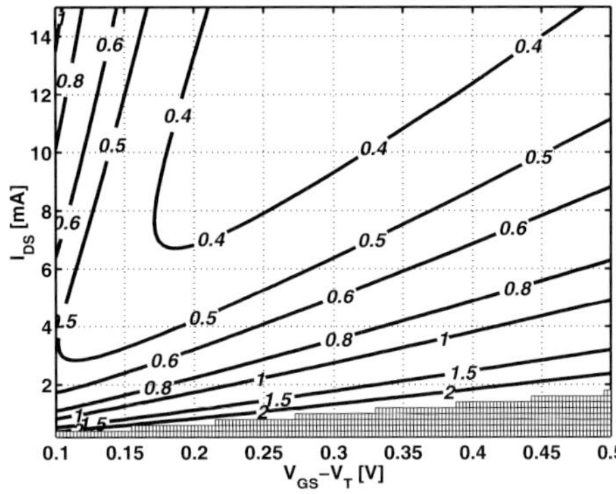

(b) Other noise contributions: $F - F_{L_g}$.

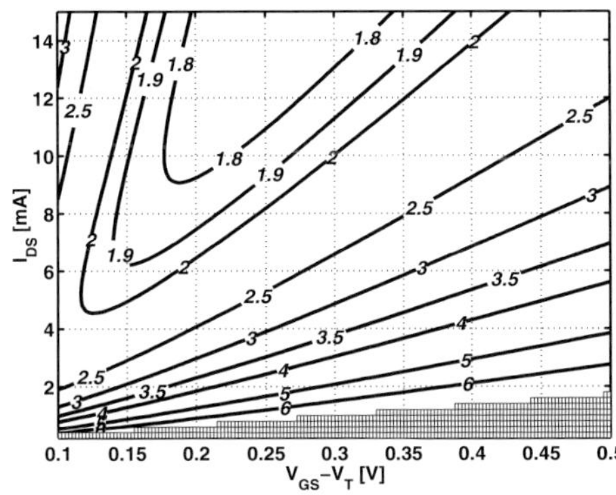

(c) Total noise figure [dB].

Figure 5.8: *Noise Contour lines of a 5 GHz* Π*-matched LNA in the design space of M1 where* $R_{S,eq} = R_S$, $R_{S,int} = 0.8R_S$ *and* $\alpha_{ind} = 2\ \Omega/nH$.

the following contribution to the noise factor:

$$F_{L_g} - 1 = \frac{R_{g,s}}{R_{S,int}}, \tag{5.14}$$

where

$$R_{g,s} = \alpha_{ind} L_g, \tag{5.15}$$

and α_{ind} is in the order of 0.5 to 2 Ω/nH depending on the technology. The gate inductor itself is found by

$$L_g = L_{g1} + L_{g2} = \frac{\sqrt{R_{S,int}\left(R_S - R_{S,int}\right)}}{\omega_0} + \frac{\sqrt{R_{S,int}\left(R_{S,p} - R_{S,int}\right)}}{\omega_0}, \tag{5.16}$$

where $R_{S,p}$ is the equivalent parallel resistance of the series connection of $R_{S,int}$ and L_{g2} or $R_{S,eq}$ and $L_{g,eq}$:

$$R_{S,p} = \frac{\omega_0^2 L_{g2}^2}{R_{S,int}} + R_{S,int} = \frac{\omega_0^2 L_{g,eq}^2}{R_{S,eq}} + R_{S,eq}. \tag{5.17}$$

The equivalent gate inductance $L_{g,eq}$ is still calculated by (4.3) and (4.4) where R_S is replaced by $R_{S,eq}$:

$$L_{g,eq} = \frac{1}{\omega_0^2 C_{gs}} - \frac{R_{S,eq} - r_{g,NQS}}{\omega_T}. \tag{5.18}$$

Consequently, $R_{S,p}$ is only dependent on the chosen value of $R_{S,eq}$ and the design of M1. Combining equation (5.14) to (5.16) allows expression of the noise contribution of the gate inductor as a function of the different equivalent resistance levels:

$$F_{L_g} - 1 = \alpha_{ind} \frac{\sqrt{R_S - R_{S,int}} + \sqrt{R_{S,p} - R_{S,int}}}{\omega_0 \sqrt{R_{S,int}}}. \tag{5.19}$$

So the higher $R_{S,int}$ the lower this noise factor contribution of the gate inductor. At higher frequencies the required inductance decreases and so does the corresponding noise contribution. These derivations also show that even though L_g and $R_{g,s}$ increase for larger $R_{S,int}$ they always increase sublinearly. Moreover, since smaller inductors have a smaller parasitic capacitance, this will further drive the design towards a high $R_{S,int}$. The limit is set by the amount of parasitic capacitance required for the ESD-protection plus the parasitic capacitance of the integrated inductor. Since in practice $R_{S,int}$ is always close to R_S, the real importance of $F_{L_g} - 1$ in the total noise figure depends mainly on $R_{S,eq}$. If $R_{S,eq}$ is sufficiently close to R_S and for sufficiently high frequencies ($\sim$ 2 GHz) this noise factor contribution can be neglected since the required inductance is then low enough.

Fig. 5.8 plots the contours for a 5 GHz LNA where $R_{S,eq} = R_S$, $R_{S,int} = 0.8 R_S$ and $\alpha_{ind} = 2\ \Omega/nH$. This value of $R_{S,int}$ decreases the bandwidth by less than 10%. The available ESD capacitance is larger than 200 fF throughout the design space. No similar plots can be shown for the L-match topology since there exists no real solution with this topology at 5 GHz with this amount of C_p. For the Π-match, the noise factor contribution due to the gate inductor is shown in Fig. 5.8(a). It decreases to the upper left. Indeed this region corresponds to a large M1 and hence a low $L_{g,eq}$. For a fixed $R_{S,eq}$ this corresponds to a smaller gate inductor L_g which lowers its noise contribution. Fig. 5.8(b) shows the sum of the other noise contributions. The transition of dominance of the classical drain noise to the gate noise can be distinguished quite clearly. To the upper left, the noise factor is more determined by the classical drain noise. To the lower right it is determined by the NQS noise contribution. The total noise figure plotted in Fig. 5.8(c) shows more or less the same behavior since the noise of L_g is significantly lower. The equation governing the noise factor behavior is listed in Table 5.3.

No plots are shown for the gain and IIP3 since the behavior is completely identical to what has been described thoroughly in Chapter 4. The only difference is the different value of $R_{S,eq}$ which can now be chosen freely. The result of a different equivalent source resistance was discussed in Chapter 4 as well. The main design equations are listed in Table 5.3. These equations are valid if

Performance parameter	Design equation
F [ ]	$1+\alpha_{ind}\frac{\sqrt{R_S-R_{S,int}}+\sqrt{R_{S,p}-R_{S,int}}}{\omega_0\sqrt{R_{S,int}}}$ $+\frac{\gamma(1+M\alpha_{gd})^2}{\alpha}\left(\frac{\omega_0}{\omega_T}\right)^2\left(g_m R_{S,eq}+\frac{2}{\kappa}\right)+\frac{\alpha\delta(1-\lvert c\rvert^2)}{\kappa g_m R_{S,eq}}+G_T^{-1}$
G_T [ ]	$\frac{R_L}{4R_{S,eq}}\left(\frac{\omega_T}{\omega_0}\right)^2\frac{1}{(1+M\alpha_{gd})^2}$
Q_{in} [ ]	$\frac{1}{2\omega_0 C_{gs}(1+M\alpha_{gd})\sqrt{R_{S,eq}R_S}}$
IIP3 [dBm]	$5.25+10\log\left(\frac{V_{GST}(2+\Theta V_{GST})(1+\Theta V_{GST})^2}{\Theta}\right)-20\log(Q_{in})$

Table 5.3: *Main design equations for the common source LNA with a* Π*-type input match.*

the Π-type matching network is sufficiently lossless. This means that all parasitic resistances are small enough such that no significant percentage of the power is lost in them. As a final remark, note that $R_{S,int}=R_S$ if $C_1=0$. Thereto, L_g needs to be implemented as an external (bonding wire) inductor. In that case $\alpha_{ind}\approx 0$ and the formulas in Table 5.3 reduce to these in Table 5.2.

5.4 Inductive ESD-Protection

In Section 5.1 the ESD-protection issue at the input of the LNA was introduced as a problem of separation. The previous topologies based on an L-match and Π-match are both focussing on the high-voltage trigger characteristics of the ESD devices. In this section the separation is done based on the difference in frequency between the RF-signal and the ESD-signal. An inductor is used to short the low-frequency ESD currents to ground while it resonates with the inherently present parasitic input capacitance C_p in order to be invisible to the RF-signal. A schematic of the proposed amplifier input is shown in Fig. 5.9(a). The integrated ESD-protection inductor, L_{ESD}, is depicted with its finite series resistance $R_{s,ESD}$. Coupling capacitor C_c couples the RF signals to the circuit. As a result the RF signal sees a high-pass filter consisting of L_{ESD} and C_c. The low-frequency ESD currents see a low-pass filter towards ground. The LC-filter acts as a signal splitter.

First take a look at this circuit from an ESD-perspective. The inductor provides a fully bidirectional path for the ESD-currents towards ground. Any bidirectional supply clamp is sufficient to have a complete protection for all ESD polarities and pin combinations. At the ESD-frequencies the impedance of the inductor is dominated by the resistance $R_{s,ESD}$ which is in fact the virtual *'on-resistance'* of the ESD device. Clearly this resistance needs to be minimized from an ESD standpoint.

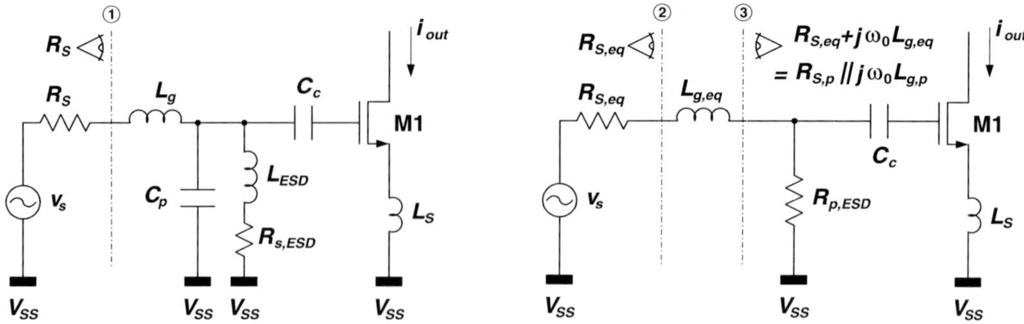

(a) Real input schematic. (b) Equivalent schematic valid @ ω_0.

Figure 5.9: *Simplified schematic of the LNA input with ESD inductor* L_{ESD}.

From RF point of view, the circuit has about the same advantage as the Π-match topology. The addition of an extra component compared to the L-match has again introduced an extra degree of freedom which can be used to optimize the performance. More concretely, any part of C_p can be tuned out in order to choose the optimum equivalent source resistance. Where a normal ESD-protection structure (cf. L-match topology) would add parasitic capacitance to the input, in this schematic, the inductor is able to tune out any or all parasitic input capacitance. $R_{S,eq}$ can again be chosen freely.

The only remaining parasitic at the input is the equivalent parallel resistance, $R_{p,ESD}$, of the parallel resonant ESD inductor. It is given by

$$R_{p,ESD} \approx \frac{\omega_0^2 L_{ESD}^2}{R_{s,ESD}} \tag{5.20}$$

and is shown graphically in Fig. 5.9(b). The impact on the performance is identical to that of the finite Q of C_p, discussed in Section 4.3.3. $R_{p,ESD}$ can be neglected in the input matching conditions if it is much larger than the equivalent parallel source resistance $R_{S,p}$ seen to the right of reference plane ③ in Fig. 5.9(b):

$$R_{p,ESD} \gg R_{S,p}, \tag{5.21}$$

where $R_{S,p}$ is given by (4.20). The left term of (5.21) is in the order of 1 to 10 kΩ while the right term is typically in the order of 100 Ω. Therefore this condition is usually fulfilled. Equation (4.20) has shown that $R_{S,p}$ reduces for smaller $L_{g,eq}$ and hence, the influence of $R_{p,ESD}$ will be reduced. Sadly, smaller $L_{g,eq}$ correspond to a larger M1 and hence higher power (for the same $V_{GS} - V_T$). This effect will pull the optimum noise performance for a given power to lower $V_{GS} - V_T$.

Now if (5.21) is fulfilled then the input matching conditions of the LNA are unchanged and the power gain remains practically unaffected. The main effect on the noise performance is an

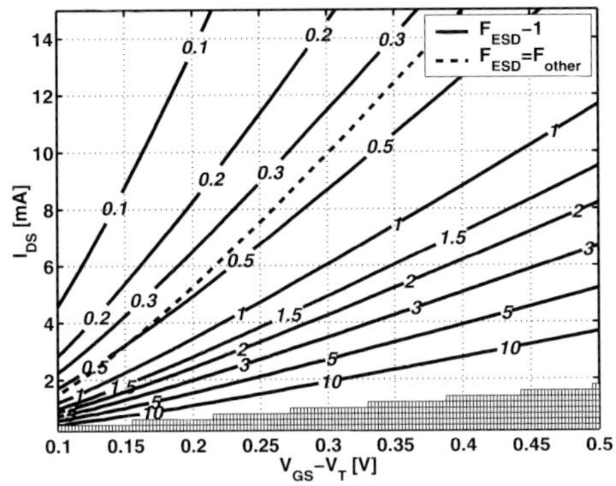

(a) Noise due to ESD inductor: $F_{ESD} - 1$.

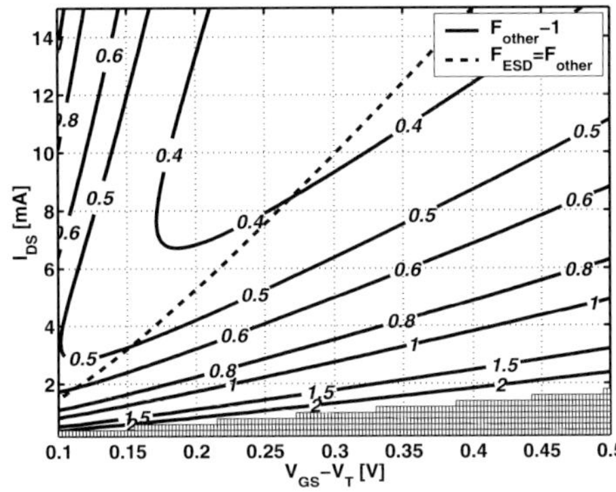

(b) Other noise contributions: $F - F_{ESD}$.

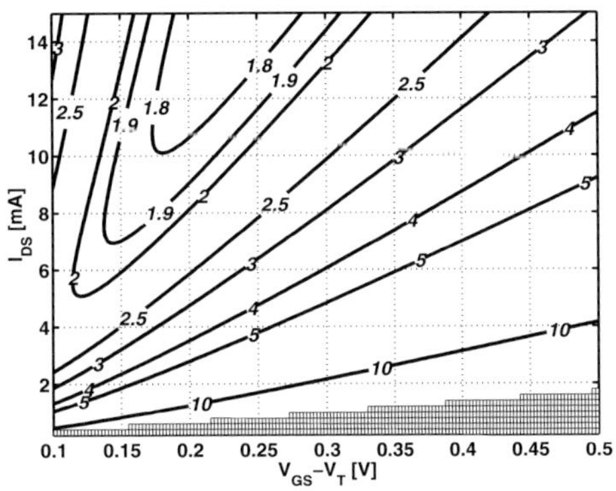

(c) Total noise figure [dB].

Figure 5.10: *Noise contour lines of a 5 GHz LNA with ESD inductor in the design space of M1 where* $R_{S,eq} = R_S$, $L_{ESD} = 2\ nH$ *and* $\alpha_{ind} = 2\ \Omega/nH$.

extra term added to the noise factor:

$$F_{ESD} - 1 \approx \frac{R_{S,p}}{R_{p,ESD}} = \frac{R_{S,eq}}{R_{p,ESD}} \left(1 + \frac{\omega_0^2 L_{g,eq}^2}{R_{S,eq}^2}\right). \tag{5.22}$$

This shows that as $R_{p,ESD}$ decreases, the noise figure will be the first to change noticeably. Naturally, the goal is to keep this noise contribution as low as possible. Hence also from an RF perspective, $R_{s,ESD}$ should be minimized. Another option is to reduce $R_{S,eq}$. If the noise of the ESD inductor is dominant $R_{S,eq}$ can be set to $R_S = 50\ \Omega$. This means the inductor tunes out C_p completely.

Fig. 5.10 demonstrates the noise behavior and bandwidth of a 5 GHz LNA with a 2 nH ESD inductor. C_p is completely tuned out and consequently $R_{S,eq} = R_S$. For the ESD inductor $\alpha_{ind} = 2\ \Omega/nH$ which is the same value as for the gate inductor L_g in the Π-match example.

Performance parameter	Design equation
F [ ]	$1 + \frac{R_{S,p}}{R_{p,ESD}} + \frac{\gamma(1+M\alpha_{gd})^2}{\alpha}\left(\frac{\omega_0}{\omega_T}\right)^2\left(g_m R_{S,eq} + \frac{2}{\kappa}\right) + \frac{\alpha\delta(1-\lvert c\rvert^2)}{\kappa g_m R_{S,eq}} + G_T^{-1}$
G_T [ ]	$\frac{R_L}{4R_{S,eq}}\left(\frac{\omega_T}{\omega_0}\right)^2\frac{1}{(1+M\alpha_{gd})^2}$
Q_{in} [ ]	$\frac{1}{2\omega_0 C_{gs}(1+M\alpha_{gd})\sqrt{R_{S,eq}R_S}}$
IIP3 [dBm]	$5.25 + 10\log\left(\frac{V_{GST}(2+\Theta V_{GST})(1+\Theta V_{GST})^2}{\Theta}\right) - 20\log(Q_{in})$

Table 5.4: *Main design equations for the common source LNA with input ESD inductor.*

This allows a more accurate comparison. These values result in $R_{p,ESD} = 1$ kΩ. Fig. 5.10(a) shows the noise factor contribution of the ESD inductor described by (5.22). Similarly to the gate inductor contribution for the Π-match LNA, this noise factor again decreases towards the upper left, for larger M1 or smaller $L_{g,eq}$. This time, however, the noise factor is directly proportional to $R_{S,p}$ where it was under the square root for the Π-match. Consequently the contours of $F_{ESD} - 1$ are closer together or in other words the behavior is more steep. Fig. 5.10(b) plots the sum of the other noise factor contributions $F_{other} - 1 = F - F_{ESD}$ where F is given in Table 5.4. These contributions are equal to the Π-match since they only depend on $R_{S,eq}$ which is equal to R_S for both amplifiers. The dashed line in both Fig. 5.10(a) and Fig. 5.10(b) indicates where $F_{ESD} = F_{other}$. For lower power consumption and/or larger $V_{GS} - V_T$, the noise of the ESD inductor is dominant. The line will shift upward if $R_{p,ESD}$ is lower. These plots also show that for a given power consumption, the best noise figures are obtained at a relatively low $V_{GS} - V_T$. However, a lower $V_{GS} - V_T$ will reduce ω_T and consequently lower the gain and IIP3 as seen from the respective equations in Table 5.4. Moreover the modelling of the transistor behavior at lower $V_{GS} - V_T$ is increasingly unreliable. For these reasons it is not advised to set $V_{GS} - V_T$ lower than 0.15 V.

5.5 Comparison

In order to be able to choose the most appropriate topology for a given application, this section aims to give a crude but nevertheless significant comparison. Remember that the main reason for the search for a new topology was that the classical L-type network is no longer able to create an input match when the parasitic input capacitance is too large. At higher frequencies the tolerated capacitance drops according to (4.30). Consequently including the ESD-capacitance and beyond a few GHz the L-match topology no longer works. Considering the design space for the L-match topology one can distinguish three degrees of freedom: I_{DS} and $V_{GS} - V_T$ of M1

L-match topology	Π-match topology	Inductive ESD-protection
I_{DS}	I_{DS}	I_{DS}
$V_{GS} - V_T$	$V_{GS} - V_T$	$V_{GS} - V_T$
$R_{S,eq}$	$R_{S,eq}$	$R_{S,eq}$
	$R_{S,int}$	
Constraint: $C_p > C_{ESD}$	Constraint: $C_1 + C_2 > C_{ESD}$	

Table 5.5: *Degrees of freedom in the design (for fixed load resistance and output matching).*

and $R_{S,eq}$ in the input match[1]. All three can be chosen freely —within certain bounds— and all other components and parameters can be calculated from these. However when C_p is fixed to the minimum allowed value consisting of the bonding pad capacitance, the wiring capacitance and the ESD capacitance, then the value of $R_{S,eq}$ is fixed for a given design of M1 as shown for instance in Fig. 4.7. This constraint reduces the real degrees of freedom to two. Moreover it will drastically limit the design space since $R_{S,eq}$ must be real and positive, as expressed by (4.28).

For the Π-match, an additional degree of freedom is created by the intermediate source resistance $R_{S,int}$. So even if the total capacitance $C_1 + C_2$ is set to the minimum allowed value, $R_{S,eq}$ can still be chosen freely for a fixed I_{DS} and $V_{GS} - V_T$. This choice will however set the value of $R_{S,int}$ and determine the corresponding bandwidth. Note also that not all combinations are allowed due to the constraint set by (4.26). However this constraint is a lot less severe than (4.28) which is not applicable here since $R_{S,eq}$ can be chosen freely. In conclusion, the Π-match offers an additional degree of freedom. The net number of degrees of freedom considering $C_1 + C_2$ is fixed to the minimum, is still three, one more than the L-match.

For the parallel ESD inductor, the reasoning is a little different. Part or all of the capacitance C_p is tuned out by L_{ESD}. $R_{S,eq}$ can again be chosen freely since for any I_{DS} and $V_{GS} - V_T$, L_{ESD} can be set specifically to yield the required value of $R_{S,eq}$. The design space is again bounded only by (4.26). The minimum capacitance constraint is no longer meaningful since any capacitance can be tuned out with the ESD-inductor. The net number of degrees of freedom is again three. An overview of the number of degrees of freedom is given in Table 5.5.

Table 5.6 gives an indication of the weak and strong points of the different topologies with respect to their performance at relatively low and high frequencies indicated by '@ LF' and '@ HF'. These notations should be interpreted with respect to the f_T of the technology and the quality of the integrated passives. The first characteristic is the number of components. This number indicates the amount of devices including their parasitics if they are relevant for the design. An integrated inductor for instance requires not only an inductance, but also one or more capacitors and resistors depending on the required model for a given frequency. Naturally the number of components is the least for the L-match topology. The gate inductor, implemented as bonding wire can be characterized sufficiently accurately, solely by its inductance. For the Π-match, the gate inductor is integrated on chip which gives extra capacitive elements and resistors to model the losses. For the ESD protection inductor idem, but the capacitive loading at the ground side of the inductor needs not be considered. The degrees of freedom and design constraints were

[1]The freedom in the design of R_L and the output matching network are not considered here.

Characteristic	L-match topology	Π-match topology	Inductive ESD-protection
Number of components	++	−	+
Degrees of freedom	3	4	3
Design constraints	1	1	0
Gain @ LF	++	++	++
Noise figure @ LF	++	+−	+
Linearity @ LF	+−	+−	+−
Bandwidth @ LF	++	++	+
ESD protection @ LF	++	++	++
Gain @ HF	—	+	+
Noise Figure @ HF	—	+	+
Linearity @ HF	—	+−	+−
Bandwidth @ HF	—	++	+
ESD protection @ HF	—	++	++

Table 5.6: *Performance comparison of the different input protection topologies.*

discussed earlier. At low frequency, the power gain of the circuit can be more than sufficient for all topologies. At high frequencies as well, but the L-match topology can no longer be used. The gain needs often be limited for reasons of stability. The noise figure can be very low for the L-match topology at low frequencies. For the Π-match it is a lot higher due to the integrated inductor which is quite large at low frequencies and has a large noise contribution. For the ESD inductor, the noise figure is minorly degraded by the extra noise from the ESD-inductor. Consequently, at moderate frequencies, where the L-match no longer works, the ESD inductor topology is a more interesting solution than the Π-match topology from a noise performance perspective. The linearity of the circuit is reasonable for all topologies and sufficient for most applications, both at low and high frequencies. The bandwidth of the LNA is highest for the L-match but is usually sufficient for all topologies. The L-match only has sufficient capacitive headroom for ESD-protection at low frequencies. For the Π-match, good ESD protection is possible at both low and high frequencies. For the ESD inductor, the protection can be very good since the inductor just needs to be able to take the current at a limited voltage drop. This means the inductance needs to be sufficiently small. If necessary (at low frequencies or for a larger $R_{S,eq}$) extra capacitance —even from an additional ESD device— can be added to ensure the correct net C_p value for the required $R_{S,eq}$.

5.6 Other ESD-Protection Strategies

5.6.1 Distributed ESD-Protection

A few other ESD-protection techniques for high frequency pins have been developed by other groups. One of these is based on a highly distributed network, approaching the behavior of a transmission line [Ito02]. The ESD devices are distributed over n different nodes separated by inductor pieces as depicted in Fig. 5.11. Each device has a limited capacitance of C_{ESD}/n. The

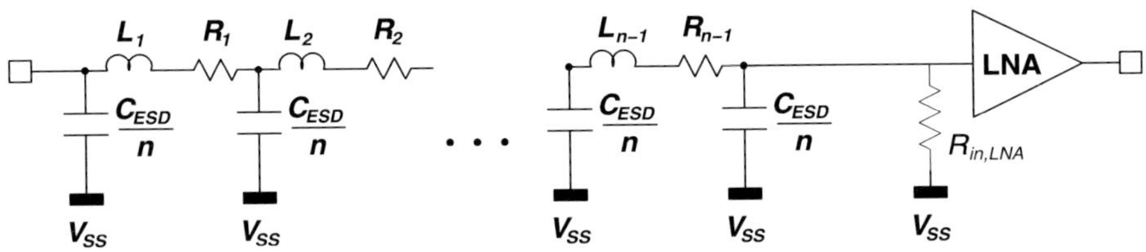

Figure 5.11: *Distributed ESD-protection for high-frequency pins [Ito02].*

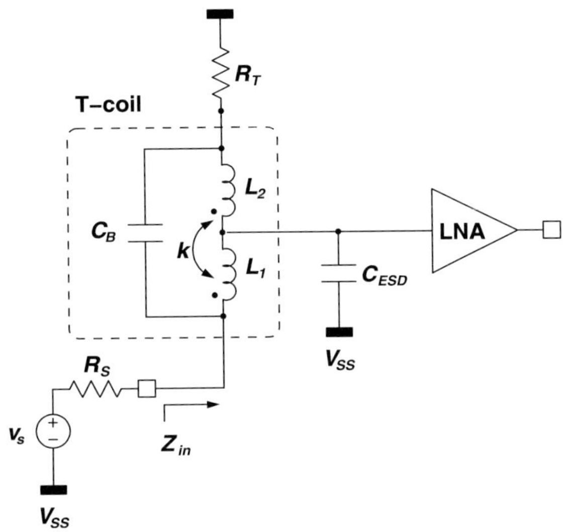

Figure 5.12: *ESD-protection for high-frequency pins with T-coils [Gal03].*

intermediate inductors can be tailored to yield the required input match. The number of devices can be increased in order to meet the bandwidth constraints. The equivalent source resistance seen by the input transistor can again be chosen freely. If the number of ESD-devices is limited to two, the circuit reduces to the Π-match topology discussed in Section 5.3.

As the number of devices is increased for bandwidth purposes, each separate ESD device becomes smaller. The ESD current flowing towards the circuit will see a higher resistance for each consecutive device since the ESD current will need to flow through the lossy inductors. This will limit the current flowing towards the input gate oxide, which is desirable. However it will largely increase the current in the early devices possibly destroying them. This can be solved by increasing the size of these early devices, which again compromises the bandwidth.

5.6.2 ESD-Protection with T-Coils

Another interesting and promising circuit uses a T-coil network to provide a broadband match regardless of the ESD capacitance [Gal03]. It can be seen intuitively that the circuit in Fig. 5.12 can match the input impedance to R_T. This means the input of the circuit can be designed to display a purely resistive input impedance, $Z_{in} = R_T$ independent of the frequency. This can be achieved independent of the value of C_{ESD}. Intuitively it can be explained as follows. At very low frequency the inductor behaves as a short and directly connects the input to the LNA. At high frequencies, the bridging capacitor performs the same function. The values of $L_1 = L_2$, C_B and the coupling factor k can be determined as function of C_{ESD} and R_T such that $Z_{in} = R_T$ independent of frequency. This topology has the huge advantage that it can be used for high-frequency baseband applications and is not limited to RF passband systems. Another advantage is that the insertion loss is low and independent of the series resistance in the inductor windings for as far as they are symmetrical in L_1 and L_2. Main drawback of the circuit is the need for a termination resistance R_T. This lower bounds the noise figure to 3 dB. A viable alternative is proposed in Chapter 7.

5.7 ESD-Protection for the Common-Gate LNA

With regards to ESD, the CG LNA is less sensitive than the CS LNA since the input is not at the gate but at the source of the transistor. As a consequence the input already has the inherent source-bulk junction which is able to carry the charges for a negative pulse vs. ground. An additional protection can be provided by the parallel inductor, used to tune out the excess input capacitance. This is basically the same strategy as in Section 5.4. Extra ESD-devices can be added, increasing the input capacitance and lowering the inductance needed to tune it out. Also the Π-network protection strategy can be used since ideally it is a lossless network and only leads to a transformation of the equivalent input impedance. The discussion is similar to Section 5.3.

5.8 Conclusion

This chapter has joined the world of RF low-noise amplifiers with the world of ESD-protection into several RF-ESD co-design methodologies. In Section 5.2 the intrinsic parasitic input capacitance has been extended with the parasitic capacitance of the ESD-protection device. Comparison of different devices has revealed that the p+ n-well diode has the ability to carry most ESD-current for a given device capacitance. It has been concluded that the protection diodes are best implemented with many separate fingers. The influence of two reverse input diodes on the overall linearity has been investigated and concluded to be negligible. Unfortunately this topology will not be available for high frequencies since the tolerated input capacitance is bounded by (4.30).

In Section 5.3 a different RF-ESD co-design concept has been proposed which does not feature this constraint. The L-match has been replaced with a Π-type matching network. The ESD capacitance is split up into two devices separated by the integrated gate inductor. This topology creates an additional degree of freedom allowing the equivalent source resistance to be chosen freely. Drawback is the integration of the inductor which adds noise to the circuit as a result of its finite quality factor. This noise is especially important when the frequency of operation is low since then the inductor will be large. This topology was deemed useful at high frequencies.

Both previous topologies use the high-current property of the ESD-pulse to separate it from the RF-signal. The circuit proposed in Section 5.4 on the other hand uses the low-frequency property of the ESD-pulse to perform the separation. An integrated inductor shunts the low frequency ESD-signal to ground. At RF-frequency the inductor tunes out any or all parasitic input capacitance. Hence the equivalent source resistance $R_{S,eq}$ can again be chosen at will. Drawback is the integrated inductor which adds noise due to its finite quality factor. Also the bandwidth is lower owing to the extra parallel input resonance. This topology yields good results at medium and high frequencies.

The three above topologies have been compared with respect to the main LNA performance criteria in Section 5.5. Section 5.6 discusses two other topologies developed in other groups. To conclude, the ESD-protection options for common-gate amplifiers have been treated briefly in Section 5.7.

Chapter 6

Integrated CMOS Low-Noise Amplifiers

6.1 Introduction

Based on the foregoing theoretical design exploration, several test chips and prototypes have been implemented in modern mainstream CMOS technologies. All circuits have been foreseen of ESD-protection. The design, layout and measurement results will be discussed. The first chip concerns a low-noise amplifier for the L2 GPS band at 1.23 GHz. It has been implemented in a 0.25 μm technology. The circuit was designed as a stand-alone amplifier, matched to 50 Ω at both input and output. A second low-noise amplifier has been designed and integrated within a complete GPS receiver front-end. It has been implemented in the same technology. This receiver focusses on the 1.57 GHz primary GPS band. The last design which will be discussed targets 5 GHz wireless LAN applications. The circuit features an integrated ESD-inductor.

6.2 A 0.8 dB NF ESD-Protected 9 mW CMOS LNA

6.2.1 The GPS Power Levels

Since two of the presented prototypes aim for application in a GPS receiver, it is useful to quickly review the signal characteristics of the GPS system. The GPS signal is broadcast at three frequencies: a primary signal at 1.575 GHz (L1 band), a secondary signal at 1.2276 GHz (L2 band) and a tertiary signal at 1.17645 GHz (L5 band) which will be introduced by the beginning of 2005. The information transmitted in these bands consists of a continuous 50 bps stream. It contains data like e.g. the satellite location, the satellite time and the necessary clock corrections. This data is spread to a much larger bandwidth by multiplication with a wide-bandwidth pseudo-random (PRN) code, commonly known as direct-sequence spread spectrum modulation (DSSS). In the receiver, the signal is de-spread by correlating it with an identical PRN sequence. Combining the received satellite data with the computed time of arrival then yields the position information.

At both L1 and L2, three different spreading codes exist:

	C/A code	P code	M code
L1	-130 dBm	-133 dBm	-128 dBm
L2	-136 dBm	-136 dBm	-128 dBm
L5	-124 dBm	—	—

Table 6.1: *Minimum GPS receive power levels.*

- The C/A code or coarse acquisition code is a code for civil use. It provides the standard positioning service. The code is short, un-encrypted and broadcast at a chipping rate of 1.023 MHz.

- The P code or the protected code is a code for military use only, providing the so-called precise positioning service. This is a much longer, encrypted code at a chipping rate of 10.23 MHz.

- The M code or military code is an extra code for military use which was launched recently. This code is optimized for anti-jamming capabilities. The chipping rate is also 10.23 MHz.

Table 6.1 shows the minimum specified received signal strength for the different GPS signals. For civil GPS, the second column (the C/A code) is the relevant one. In the L1 band (broadcast at 1.575 GHz) the minimum received power is -130 dBm. This gives an effective SNR of about 29 dB at the input of the receiver. In the L2 band (broadcast at 1.2276 GHz) , the minimum received power is even 6 dB lower, yielding an effective SNR of 23 dB. In practice, the SNR of the received signal is much worse. In urban canyons or when tree foliage shadows the user, the minimum received power often is much lower than the specified -130 dBm. The SNR can be degraded by as much as 10 to 20 dB.

Hence, to keep the receiver from failing at low input signal levels, the receiver noise figure must be very low. This poses severe demands on both the noise figure and the gain of the RF input amplifier. To cope with these requirements, often high-performance GaAs MESFET low noise amplifiers are used, since they are capable of offering excellent noise figures in the order of 1 dB at large power gains of 20 dB. In order to prove the suitability of CMOS for building extremely sensitive receivers, one must demonstrate the feasibility of achieving very low noise figures (≤ 1 dB) and large gains (18 to 20 dB) at a power consumption comparable to GaAs solutions. In [Ler01a], a CMOS LNA has been presented which consumes less than 10 mW while offering this level of performance. This design is discussed next.

6.2.2 Topology

The LNA schematic is shown in Fig. 6.1. The input of the LNA is protected against ESD by two reverse-biased diodes. The power supply lines, the gate of the cascode device (and the associated bondwires) are bypassed to ground using 40 pF decoupling capacitors. Special care has been taken to sufficiently dampen the parasitic resonances that may occur (i.e. the parallel resonances of the bondwire inductance and the decoupling capacitor).

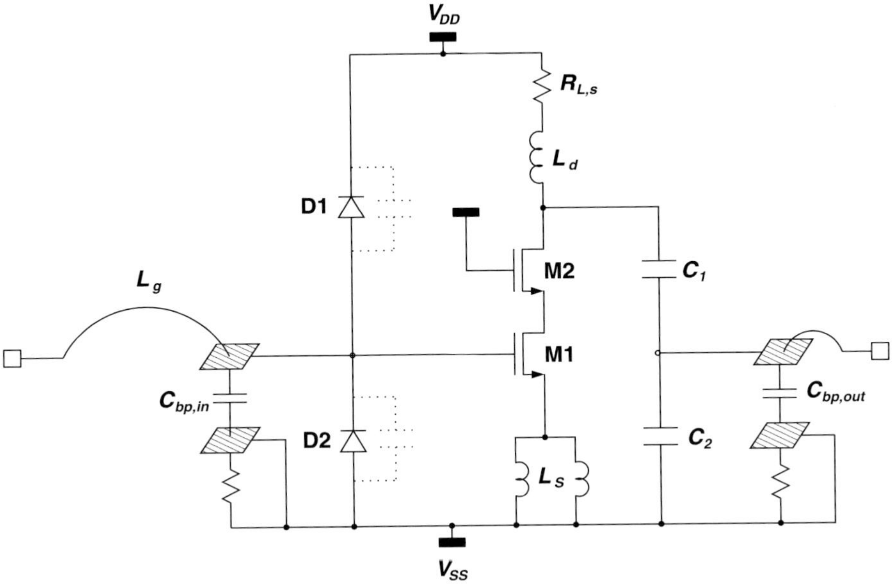

Figure 6.1: *Simplified schematic of the LNA.*

Since the LNA was designed as a stand-alone circuit with both input and output impedance equal to 50 Ω, achieving a maximum power transfer requires that the 50 Ω load is transformed into the complex conjugate of the effective output impedance at the drain of the cascode. In other words, the matching network must transform the 50 Ω load into a resistive path with impedance R_L and at the same time generate the exact amount of parallel capacitance to cancel out the effective inductance at the drain of the cascode. Therefore, the rest of the matching network must contain two degrees of freedom. In this particular circuit, these degrees of freedom are offered by the quasi-lossless capacitive divider C_1/C_2 [Flo99]. In fact, for each realizable inductor there exists a realizable combination of C_1 and C_2 values that provides the correct impedance, provided that

1. The inductor is not self-resonant at frequencies near or below the operating frequency.
2. The required capacitance from the output node to V_{SS} is larger than the minimum possible which is limited by the sum of the parasitic capacitance of the output bond pad and the stray capacitance of C_1 towards the substrate.

6.2.3 Design

The supply voltage of the circuit was set to 1.5 V and the total power consumption of the LNA was aimed below 10 mW. This corresponds to a current of 6.7 mA. In order to clarify the design trade-offs, Fig. 6.2 and Fig. 6.3 show contour plots of the most important LNA properties. These

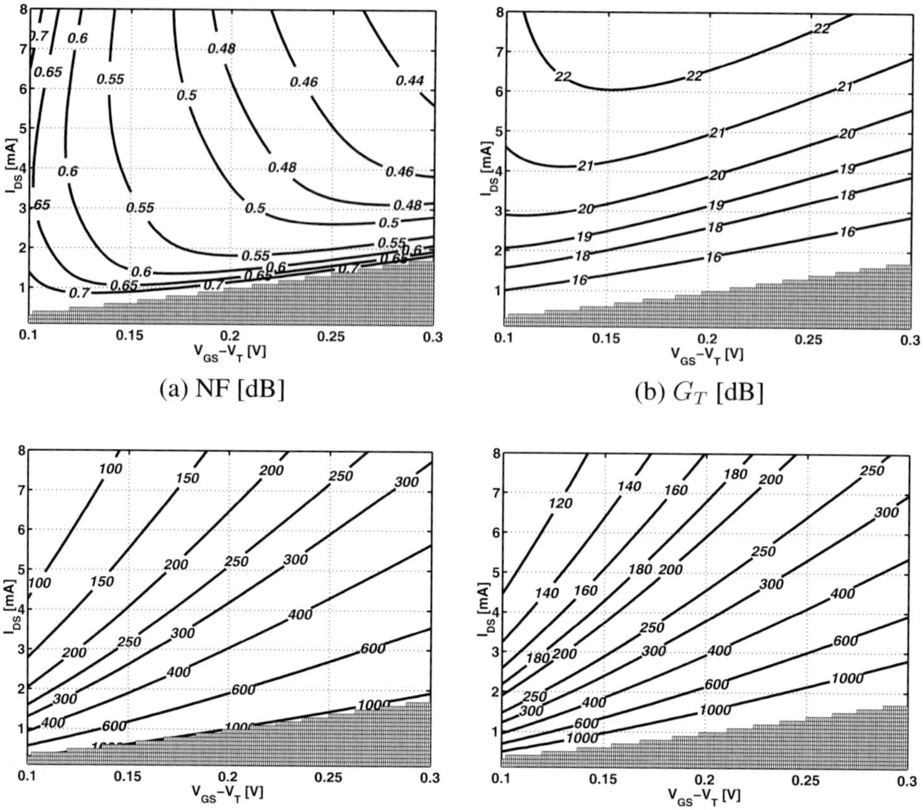

(a) NF [dB] (b) G_T [dB]

(c) $R_{S,opt}$ [Ω] (d) $R_{S,eq}$ [Ω]

Figure 6.2: *Contour lines for the 1.23 GHz LNA.*

contours are calculated based on the parameters in Table 2.1. The noise parameters γ and δ were assumed to be equal to 2 and 4 respectively. These values were based on a correlation of the simulation and measurement results. The input capacitance C_p is set to 210 fF, i.e. 110 fF for the bonding pad and 100 fF for the protection diodes. Fig. 6.2(a) depicts the noise figure of the LNA under ideal circumstances (i.e. assuming a lossless L_g, L_s, etc.). As can be seen from the plot, the noise figure is extremely low in the whole design space. The LNA doesn't even need the available 6 mA: according to the plot, a noise figure as low as 0.6 dB can already be achieved at a drain current of only 1.5 mA. Fig. 6.2(c) plots the optimum source resistance for each point in the design space. Comparison with the actual equivalent source resistance corresponding to the C_p of 210 fF and plotted in Fig. 6.2(d), shows that they are relatively close in a large region. Both increase towards the lower right of the graph.

Fig. 6.2(b) indicates that the power gain drops for a fixed R_L when biasing the input stage at low current levels. This can be attributed to the drop in the efficiency of the amplifying device due to the increase in the equivalent source resistance, $R_{s,eq}$ seen by the input transistor. The gain

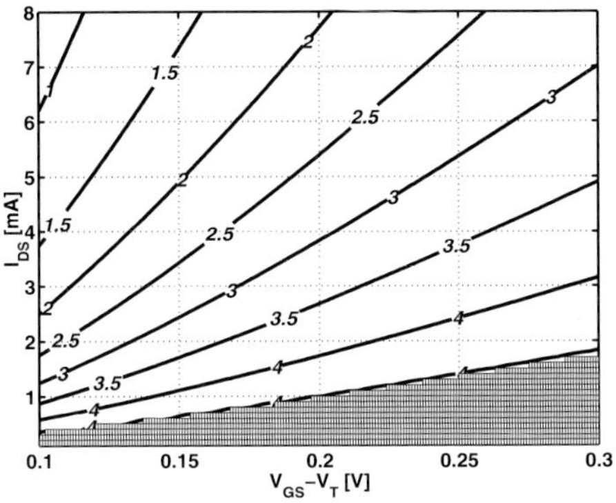

(a) Q_{in} [ ].

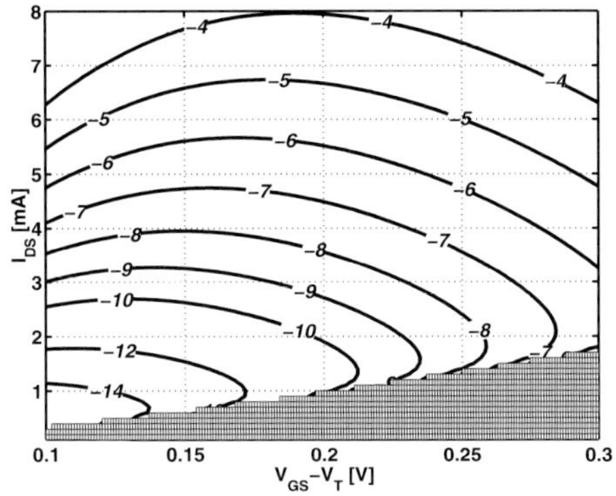

(b) IIP3 due to M1 [dBm].

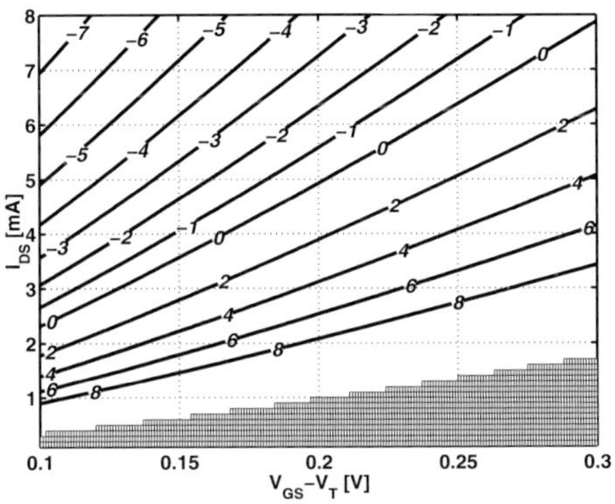

(c) IIP3 due to C_{db2} [dBm].

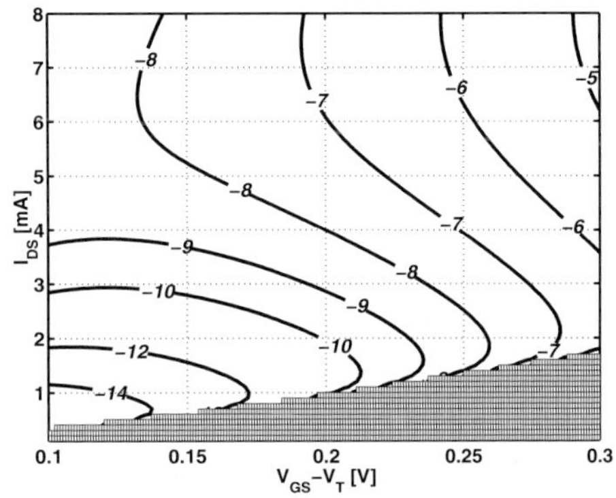

(d) Total IIP3 [dBm].

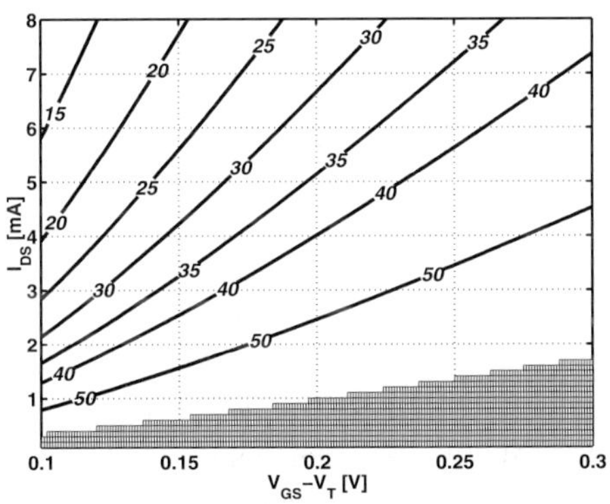

(e) L_g [nH]

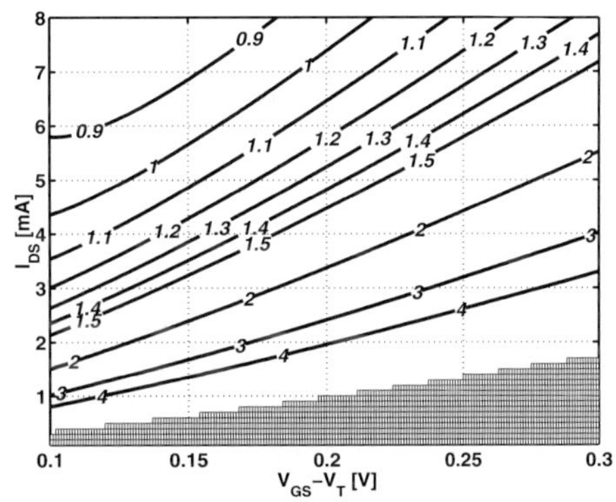

(f) L_s [nH]

Figure 6.3: *Contour lines for the 1.23 GHz LNA.*

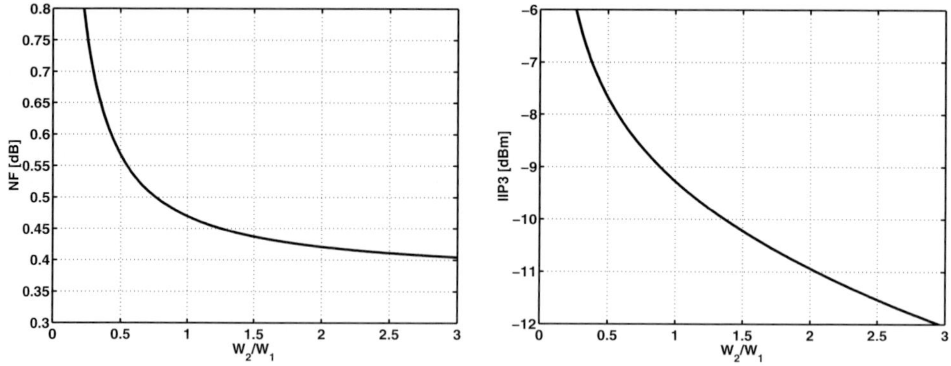

(a) NF [dB] (b) IIP3 [dBm]

Figure 6.4: *Noise figure and IIP3 as a function of the relative width of M2 with respect to M1.*

could only be increased by increasing the load resistance. However, practical values of R_L are limited to a few hundred Ohm due to parasitic load contributions and considering reliability with respect to process variations. Both effects have been explained in Section 4.8. Furthermore, an inductor with a large R_L must necessarily exhibit a relatively large inductance value due to the maximum available Q factor. Considering that the resonance frequency must remain the same, this strongly limits C_1 and C_2, which makes the matching network very sensitive to external parasitics. For instance, for $\alpha_{ind} = 1$ Ω/nH the maximum Q is $\frac{\omega_0}{\alpha_{ind}} \approx 8$. In that case, an R_L of 300 Ω corresponds to a minimal inductance of $\frac{R_L}{Q_{L,max}\omega_0} \approx 5$ nH. Often α_{ind} is even higher in order to limit the parasitic capacitance of the inductor. This makes even less capacitance available for the divider. Consequently, the design of the inductor is based on maximizing the load resistance while keeping sufficient capacitive headroom for the divider. This design strategy yielded a 10.5 nH inductor with an R_L of 330 Ω or a series resistance of 20 Ω. The inductor uses only the top metal layer in order to minimize the parasitic capacitance. Capacitors C_1 and C_2 are in the range of 1 pF. The exact design values can be found in Table 6.2.

Fig. 6.3(b) plots the contour lines of the IIP3 of the LNA. Clearly, 6 mA is not required from a linearity perspective as well. An IIP3 of -10 dBm can already be obtained at about 3 mA. However since transistor distortion simulations are rather inaccurate, a few dB of margin is welcome. Cascode transistor M2 was chosen somewhat smaller than M1. This yields a more pronounced Miller effect which increases the noise figure and reduces the cascode pole. However, both effects are minor and do not outweigh the benefits of the lower capacitive loading at the drain of M2. This helps to free capacitance for the divider. And more importantly it improves the overall linearity by reducing the contribution of the non-linear drain-bulk capacitance of M2. This IIP3 contribution is given by (4.66) and plotted in Fig. 6.3(c). It is assumed in these plots that the width of M2 is 5/9 of the width of M1 (this value is explained in the next paragraph). The IIP3 solely due to C_{db2} decreases very rapidly towards the upper left. These design points are characterized by a small $R_{S,eq}$ and a large M1 and hence, a large C_{db2}. Both effects reduce the IIP3

M1	I_{DS}	6 mA	M2	I_{DS}	6 mA
	$V_{GS} - V_T$	0.14 V		$V_{GS} - V_T$	0.2 V
	W/L	450/0.25		W/L	250/0.25
	g_m	64 mS		g_m	46 mS
	C_{gs}	320 fF		C_{gs}	190 fF
Gate inductance	L_g	24 nH	Load inductor	L_d	10.5 nH
Source inductance	L_s	1 nH		$R_{L,s}$	20 Ω
ESD diodes	A_D	2x37 μm^2		$R_{L,L}$	330 Ω
	P_D	2x25 μm	Capacitive divider	C_1	0.96 pF
	C_D	2x50 fF		C_2	1.16 pF
	R_D	~ 6 Ω	RF bondpads	C_{bp}	110 fF

Table 6.2: *Values for the main design parameters and components. Technology parameters can be found in Table 2.1.*

as seen from (4.66). This equation also shows that a double width for M2 would decrease this IIP3 contribution by an additional 3 dB. The total IIP3 considering both the non-linearity of M1 and C_{db2} is indicated in Fig. 6.3(d). Towards the bottom right, it is mainly determined by M1. Further to the upper left it is mainly determined by C_{db2}.

Fig. 6.3(e) shows the required gate inductance in order to obtain a purely resistive input impedance. This is one of the main reasons to focus the design in the upper left of the graph. There, the gate inductors are lower due to the large width of the input transistor. In view of the above discussion, the I_{DS} and $V_{GS} - V_T$ of the amplifying device have been set to 6 mA and 0.14 V, respectively. The corresponding gate inductance is 24 nH. Fig. 6.4 plots the noise figure and IIP3 as a function of the relative width of cascode transistor M2. Both increase rapidly for smaller M2. The ratio has been set to 5/9. Compared to a 1/1 ratio this yields a noise figure degradation of 0.07 dB owing to the increased Miller effect. It is accompanied by an IIP3 improvement of 1.3 dB. The drain-bulk capacitance of M2 is reduced from 180 fF to 100 fF. Component values and transistor sizes are listed in Table 6.2.

6.2.4 Layout

The analysis in the previous section assumes that all passive components — except for L_d — are lossless and hence, noiseless. The layout must ensure that this is effectively the case. How this is done is indicated in this section.

transistors Both the amplifying device and the cascode device employ a finger structure. The gate resistance has been reduced as much as possible by using short fingers (5/0.25). Since these short fingers allow the distance between the bulk contacts and the middle of the channel to be lower, the effective substrate resistance can be decreased so that less noise is injected through the back gate. The bulk contacts also shield it from injected output signals, cutting a possibly dangerous feedback loop.

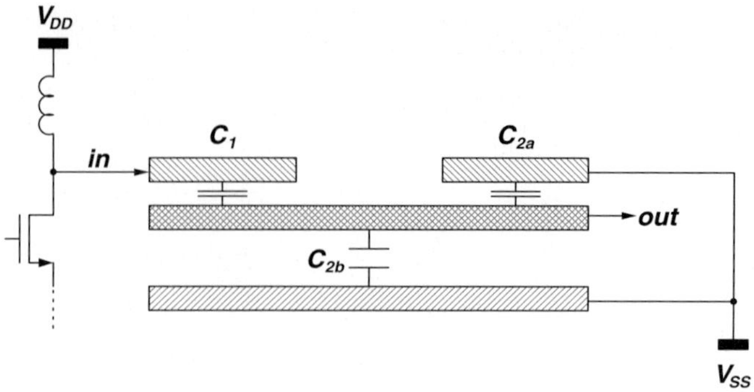

Figure 6.5: *Cross-section of the capacitive divider in the output matching network.*

bonding pads To prevent the substrate resistance from injecting noise currents, the input pad is shielded from the substrate by a grounded metal layer underneath the bondpad. A detailed explanation of its function can be found in Section 4.10.1.

ESD diodes The diffusions connected to the supply lines completely enclose the diffusion regions connected to the RF input terminal to reduce the series resistance of the two-diode protection as much as possible. Evidently, the spacing between the p and n diffusions needs to be minimized since we are dealing with high-ohmic substrate material.

load inductor The load inductor is implemented in the fourth and the third metal layer, minimizing its parasitic capacitance towards the substrate. A patterned ground shield underneath the inductor shields the inductor from the lossy substrate by providing a low-ohmic path to ground as discussed in Section 4.10.2.2. The ground shield is implemented in the first metal layer.

output capacitors The capacitors of the output matching network (C_1 and C_2) are implemented as metal–insulator–metal (MIM) capacitors featuring a very low series resistance. The stray capacitance to the substrate is intercepted by a ground shield and also contributes to C_2 (see Fig. 6.5). The ground shield has originally been inserted to bypass the substrate resistance, improving the quality factor of the output network. The ground shields in this layout — the shields underneath the pads and the output capacitors, the patterned ground shield underneath the inductor — also serve another purpose: increasing the on-chip reverse isolation. Ultimately, the reverse isolation is believed to be limited by the cross-talk between the external bondwires.

high level layout The power supply lines, the gate of the cascode device are bypassed to ground using 40 pF decoupling capacitors. Special care has been taken to sufficiently dampen the parasitic parallel resonance with the associated bonding wires. The IC is implemented in a standard 0.25 μm 4M1P CMOS process and occupies an area of 0.66 mm^2. A photograph of the IC is shown in Fig. 6.6.

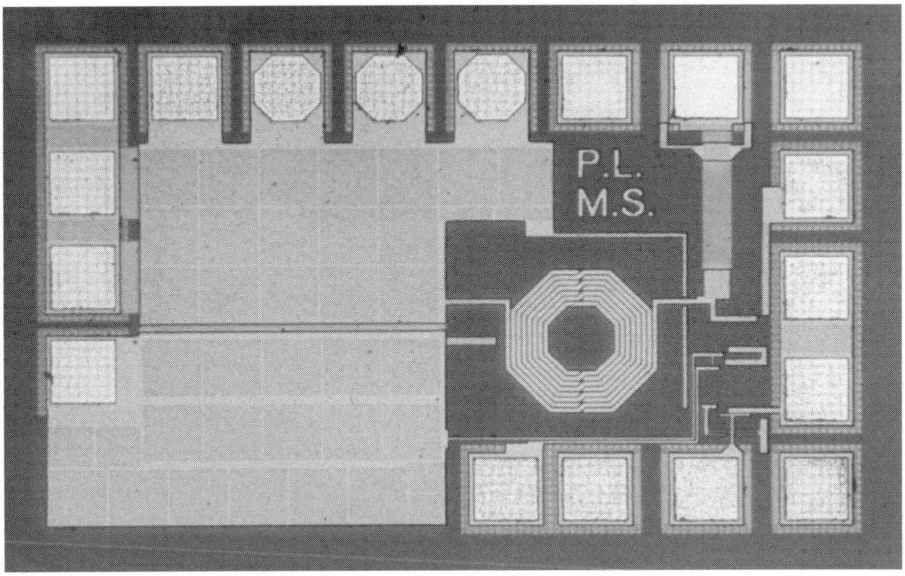

Figure 6.6: *Micrograph of the 1.23 GHz LNA.*

6.2.5 Experimental Results

For the measurements the IC is glued onto a thick film ceramic substrate and all the pads are wire bonded to 50 Ω microstrip-lines. The gate inductor is implemented as a bondwire because of its low series resistance and its low parasitic capacitance. The substrate is then mounted in a Copper-Beryllium box which shields the LNA from external interference and serves as reference ground. The connection to the external world is provided through two SMA connectors.

The LNA is biased in its nominal 9 mW regime, i.e. drawing 6 mA from a 1.5 V supply. The complete S-parameter set has been measured using an HP network analyzer. The power gain, S_{21}, plotted in Fig. 6.7(a), is measured to be a flat 20 dB in a 100 MHz wide band around the GPS L2 frequency of 1.2276 GHz (1.2–1.3 GHz). The -3 dB bandwidth is approximately 400 MHz (1.05–1.45 GHz). At the same time, the reverse isolation ($-S_{12}$) is better than 31 dB over the whole frequency range of the network analyzer (300 kHz–3 GHz). Fig. 6.7(b) shows that, within the L2-band, the input reflection coefficient (S_{11}) and the output reflection coefficient (S_{22}) are -11 dB and -11.5 dB, respectively. Both reflection coefficients are better than -10 dB in a 100 MHz wide band around the GPS L2 frequency of 1.2276 GHz (1.2–1.3 GHz). Due to the increased resistivity of the top metal layer, the R_L of the coil had become 20 percent lower than originally simulated, which resulted in a lower S_{21} and a larger S_{22}. The gain degradation has been compensated for by lowering the input impedance to 30 Ω by decreasing the nominal L_s value. The noise figure of the 50 Ω R_S/30 Ω R_{in}-configuration is approximately the same as in case of a normal 50Ω R_S/50Ω R_{in}-configuration. The noise figure of the LNA has been measured directly using a noise figure meter and is plotted in Fig. 6.7(c). At the GPS L2 frequency, a low

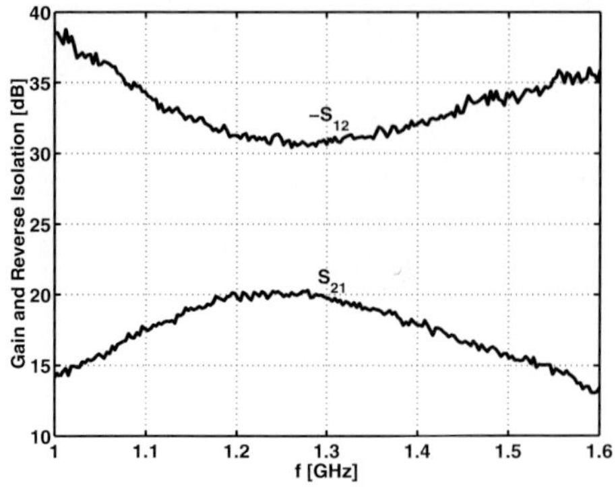

(a) Gain and reverse isolation [dB].

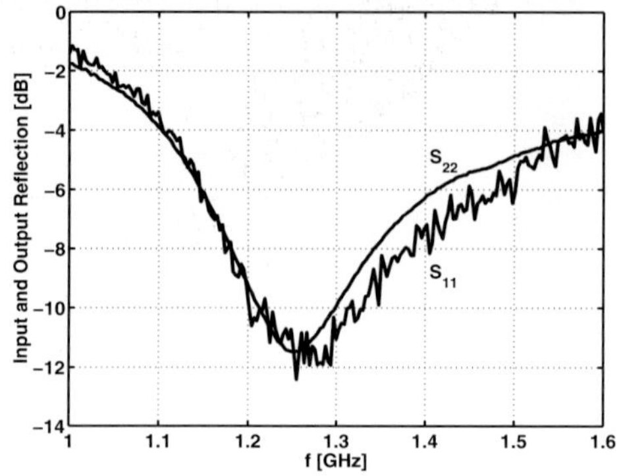

(b) Input and output reflection [dB].

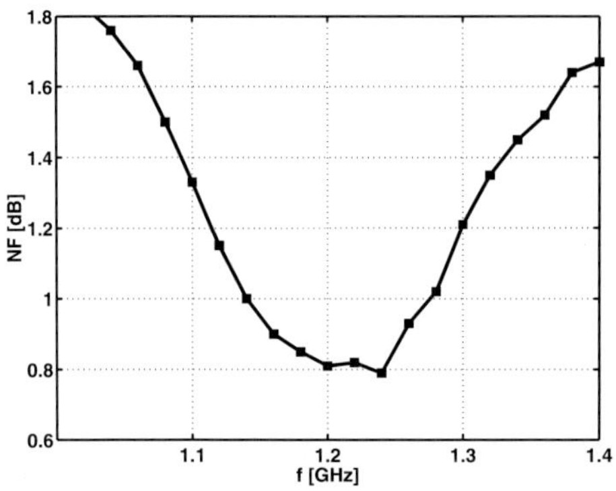

(c) Noise Figure [dB].

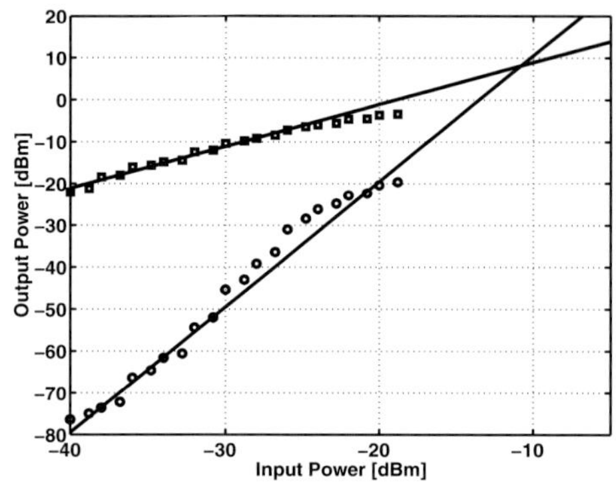

(d) Intermodulation distortion.

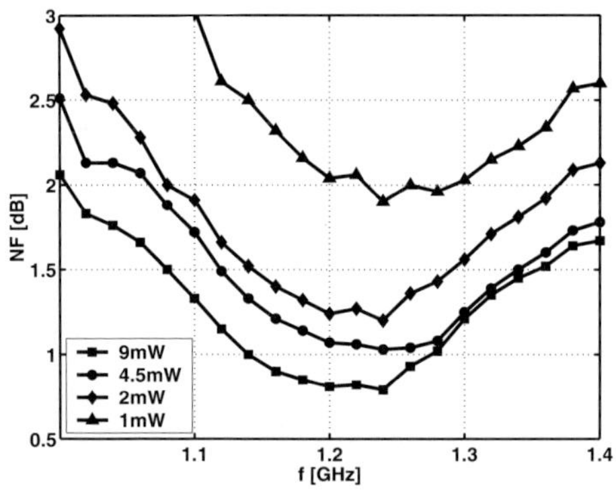

	Power	*Supply*	*Current*
■	9 mW	1.5 V	6 mA
●	4.5 mW	1.5 V	3 mA
♦	2 mW	1 V	2 mA
▲	1 mW	1 V	1 mA

(e) Noise figure for different biasing conditions.

Figure 6.7: *Measured S-parameters, noise figure and intermodulation distortion.*

Supply Voltage	1.5 V
Current consumption	6 mA
Power consumption	9 mW
NF	0.8 dB
S_{21}	20 dB
S_{11}	$\sim$ -11 dB
S_{22}	$\sim$ -11 dB
S_{12}	$<$ -30 dB
IIP3	-11 dBm
HBM voltage: IN+ vs. V_{DD}-	600 V
HBM voltage: IN- vs. V_{SS} +	-1400 V

Table 6.3: *Compilation of the main measurement results.*

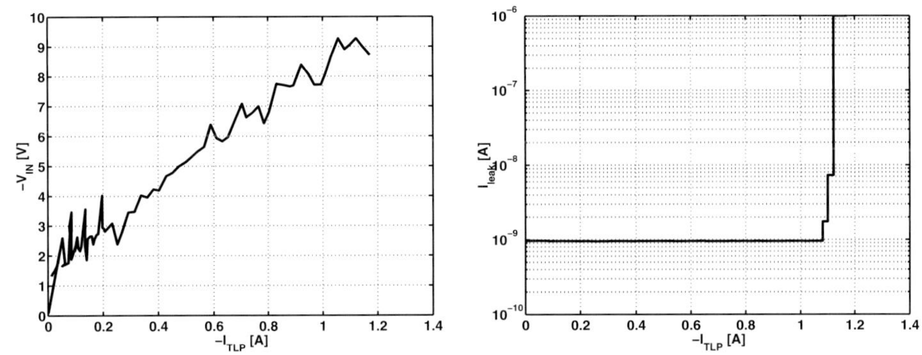

(a) Input voltage vs. ESD current. (b) Measured leakage current.

Figure 6.8: *Results of a TLP test: input versus V_{SS}.*

noise figure of 0.8 dB is measured (including the noise of the microstrip-lines). In addition, the noise figure remains below 1.2 dB in the 200 MHz wide frequency range between 1.1 GHz and 1.3 GHz. The noise figure has been measured for a few different biasing conditions. The resulting noise figures are shown in Fig. 6.7(e). It has a minimum of about 2 dB at 1 mA. Naturally, the input matching requirement is no longer fulfilled. The lower current has reduced the $V_{GS} - V_T$ of M1. Therefore the input resistance will be too low owing to the reduction in ω_T.

The sensitivity to nearby interferers has been evaluated as well. Fig. 6.7(d) shows that the input-referred third-order intercept point (IIP3) and the 1 dB compression point are -10.8 dBm and -24 dBm, respectively. The IIP3 can be correlated with the corresponding simulated value of -8 dBm found in Fig. 6.3. It is worth noting that all the measurements have been performed from SMA connector to SMA connector, i.e. without de-embedding the substrate parasitics like strip-line resistance, connector non-idealities, etc..

The IC has also been tested for ESD-immunity. A TLP test was used to estimate the maximum current the circuit can handle. The result is shown in Fig. 6.8 for a negative pulse between

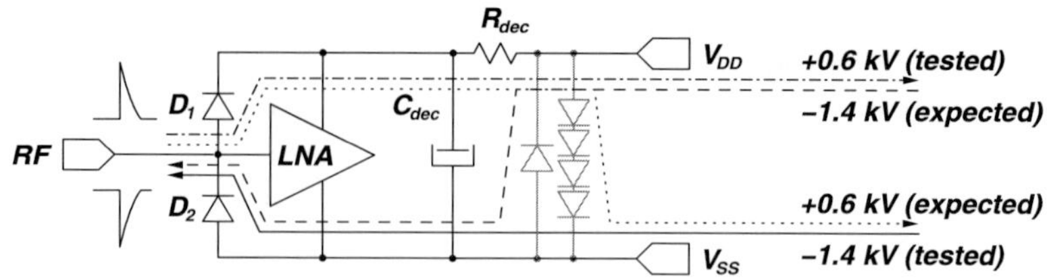

Figure 6.9: *The origin of the measured ESD performance.*

input and V_{SS}. The transmission line precharge voltage is gradually increased, yielding a larger ESD-current in each consecutive zap. After each zap the leakage current is measured at the input by applying a fixed voltage far below breakdown. This is illustrated in Fig. 6.8(b). At a certain point the leakage current drastically changes indicating that the circuit is destroyed. In this test this event occurred at an ESD-current of -1.1 A. Converting this to the corresponding HBM voltage can be done by multiplying this current with the HBM discharge resistance of 1.5 kΩ. This yields an estimated HBM protection voltage of -1.65 kV. The IV characteristic of the ESD-protection is plotted in Fig. 6.8(a). The slope of the curve gives an indication of the on-resistance of the ESD-protection. In this test it is in the vicinity of 6 Ω.

HBM ESD-tests have shown that the LNA is capable of surviving positive ESD pulses up to 0.6 kV (zaps measured with respect to V_{DD}) and negative ESD pulses down to -1.4 kV (zaps measured with respect to ground). The latter corresponds well with the 1.65 kV predicted by the TLP measurements. The origin of the lower performance for positive pulses to V_{DD} can be clarified using Fig. 6.9. The bottom diode (D2) protects the input against negative zaps with respect to ground, yielding a protection of -1.4 kV. Positive zaps with respect to V_{DD} are covered by top diode D1. However, the series resistance originally inserted in the V_{DD} path to damp any possible resonance between the power supply bondwire and the decoupling capacitors, lies in the discharge path and therefore limits the positive ESD performance to the lower 0.6 kV value. In case of a positive zap with respect to ground, the top protection diode must conduct the positive ESD current to the V_{DD} from where it must be directed to ground through a low-resistance power supply clamp. However, since this clamp was not implemented on the test chip, we could not test the susceptibility to positive ESD pulses with respect to V_{DD}. For exactly the same reason we could only test the susceptibility to negative ESD pulses with respect to ground and not with respect to the V_{DD}. Nevertheless, since such a clamp may consist of very large structures which contribute almost no series resistance to the ESD discharge path, the LNA should be able to withstand 0.6 kV positive zaps with respect to ground and -1.4 kV negative zaps with respect to V_{DD}. The measurement results are summarized in Table 6.3.

6.2.6 Discussion and Comparison

In order to be able to position this work with respect to existing LNA's, Table 6.4 lists the performance of the CMOS power-gain LNA's published to date.

The 0.8 dB noise figure offered by this LNA is the lowest noise figure ever reported in CMOS.

Ref.	f_0 [GHz]	NF [dB]	P_{DC} [mW]	G_T [dB]	IIP3 [dBm]	ESD [kV]	L_{min} [μm]	NF$_{sys}$ [dB]	η_{NF} [dB]
[Kar96]	0.9	2.2	20	15.6	-3.2	—	0.5	3.1	18.8
[Sha97b]	1.6	3.8	12	17	-6	—	0.35	4.3	17.8
[Sha97a]	1.5	3.5	30	22	-9.3	—	0.6	3.7	14.3
[Zho98a]	0.9	4.1	18	12.3	—	—	0.6	5.3	15.5
[Stu98]	0.9	2.1	24	12	+4	±1	0.5	4.1	18.3
[Hay98]	0.9	1.8	9	14.8	-2.5	—	0.35	3.0	23.4
[Hua98a]	0.9	1.74	45	14.3	-2.8	—	0.25	3.1	16.4
[Hua98b]	0.9	1.85	27	16.2	-7.2	—	0.25	2.8	18.4
[Flo99]	0.9	1.2	30	22	-1	—	0.8	2.7	20.2
[Sam99]	5.3	4.8	7.2	—	-2	—	0.25	—	18.4
[Gra00a]	0.9	1.75	27	10	+4.7	—	0.35	4.7	18.7
[Gra00b]	0.9	1.05	9	11.4	-2	—	0.35	3.7	26.0
[Gra00b]	0.9	0.9	18	13.5	0	—	0.35	2.8	23.8
[Gra00b]	0.9	1.2	13.5	12.8	-3	—	0.35	3.2	23.7
[Din01]	0.9	2.8	45	—	18	—	0.35	—	13.9
[Bru02]	0-1.6	2.2	35	—	0	—	0.25	—	16.4
[Cha02]	5.2	2.45	26.4	19.3	-6.1	—	0.35	2.9	17
[Cas03]	5.8	0.9	16	14.2	0.9	—	0.18	2.5	24.3
[Ler01a]	1.2	0.8	9	20	-10.8	-1.4/0.6	0.25	1.3	27.4

Table 6.4: *Performance summary of recently published CMOS LNA's.*

In addition, the LNA is the first sub-1 dB noise figure LNA at such a low power consumption (9 mW); The only other sub-1 dB NF LNA is the 0.9 dB noise figure LNA in [Gra00b], which consumes twice the power and features a 6 dB lower gain!

The power gain of 20 dB is the second largest gain in the table. In a system, this LNA suppresses the noise of the subsequent stages by a factor of 100, making it very suitable for applications where a low system noise figure is required. Most published CMOS LNA's that attain a low noise figure simply do *not* have sufficient gain to yield a low noise receiver. This can be shown by looking at the system noise figure, defined as

$$\mathrm{NF}_{sys} = 10\log_{10}\left(F_{LNA} + \frac{F_{mixer} - 1}{G_T}\right), \tag{6.1}$$

which is a good figure-of-performance to evaluate the capability of an LNA when it is inserted in a real system. Table 6.4 states NF$_{sys}$ for all LNA's in case the mixer noise figure amounts to 12 dB. Whereas all CMOS LNA's in the table offer system noise figures between 2.5 and 5.5 dB, the presented CMOS LNA enables total receiver noise figures as low as 1.3 dB! And, since the noise figure of CMOS mixers is usually higher than 12 dB, the difference between the presented LNA and the existing LNA's is even more pronounced in reality. Another performance number

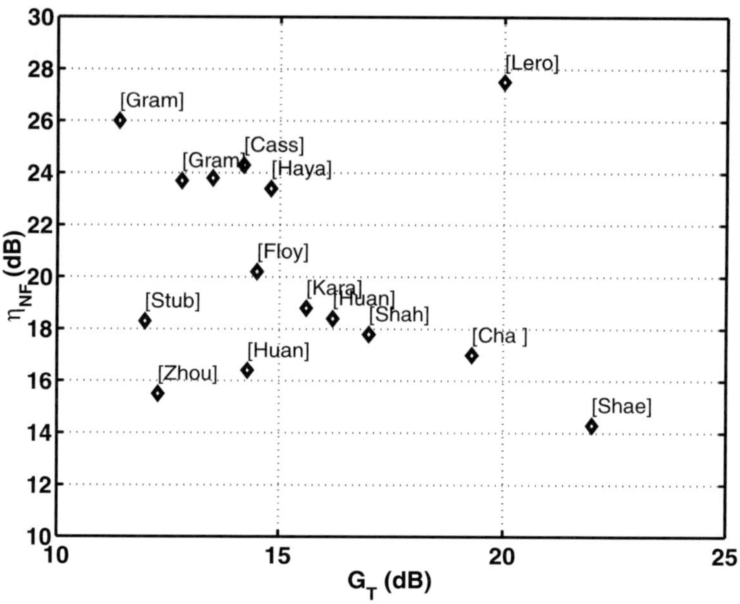

Figure 6.10: *Comparison of recently published CMOS LNA's.*

is shown in the last column of Table 6.4. It represents the power efficiency with which the LNA noise figure has been realized and is defined as

$$\eta_{NF} = -10\log_{10}(\frac{(F-1)P_{DC}}{1\ \text{W}}). \tag{6.2}$$

It is important to note that both the input and output reflection coefficient of the presented LNA (-11 dB and -11.5 dB, respectively) comply to the -10 dB filter termination requirement. The 0.8 dB noise figure is thus accompanied by an acceptable amount of reflection[1]. Moreover, all the matching networks are integrated on-chip (except for the input bondwire). Other low-power low-noise amplifiers use external input and output matching networks, and exploit these additional degrees of freedom to balance a noise match with an input match; For instance, by using an intermediate reference plane at a lower impedance value. In a solution with a single external inductance this degree of freedom is simply not available.

The linearity — although somewhat lower than the rest — is more than acceptable; The measured IIP3 value of -10.8 dBm is more than enough for the GPS application (higher than -20 dBm). In the sensitive GPS receiver, the gain of the LNA needs to be quite large in order to bring the signal sufficiently above the noise floor of the mixers. As a result, the IIP3 of the down-conversion mixer generally dominates the IIP3 of the receiver. Therefore, the IIP3 is not

[1]in contrast to some other LNA's that feature a good noise performance at the cost of a lousy input match, like e.g. [Kar96] and [Flo99].

so much a spec for an LNA as it is for a mixer; As long as the linearity of the LNA is large enough, there is no problem whatsoever.

Apart from its raw performance, one of the most important properties of this LNA is that it features some degree of protection against ESD. The IC is fitted with an ESD protection on the RF input, which is capable of protecting the LNA against -1.4 kV to 0.6 kV HBM zaps. Almost none of the published low noise amplifiers include any ESD protection.

To conclude, Fig. 6.10 shows yet another way of positioning this work with respect to previously published CMOS LNA's. The figure on the vertical axis is η_{NF}. The horizontal axis represents the LNA gain (the LNA's capability of determining the system noise figure). The more the LNA is positioned towards the upper right corner, the better its performance. The presented LNA clearly outperforms all other published CMOS LNA's in this regard. [2]

6.2.7 Conclusion

The 0.25 μm CMOS LNA described here offers a noise figure as low as 0.8 dB at a power gain of 20 dB while consuming only 9 mW. This design is competitive with current commercially available GaAs LNA solutions. In addition, the IC is fitted with an ESD protection on the RF input, which is capable of protecting the LNA from -1.4 kV to 0.6 kV HBM. This demonstrates that an excellent performance can still be achieved while at the same time providing >0.5 kV ESD protection. This design outperforms previously published CMOS LNA's with respect to noise figure, gain and power consumption.

6.3 A 1.3 dB NF CMOS LNA for GPS with 3 kV HBM ESD-Protection

6.3.1 The Complete GPS Receiver Front-End

6.3.1.1 Architecture

Many recent wireless receivers are based on the Low IF architecture. It combines the advantages of a low frequent IF and a power efficient image rejection without the need for external high Q filtering. The analog outputs of such front-ends generally consist of 2 quadrature signals (I and Q). In early quadrature low-IF receivers, both I and Q paths were digitized by two independent low- pass ADC's. This technique does not take into account the complex character of the I and Q signals when combining the two paths. Since the wanted signal band is only located at positive frequencies, the power efficiency of the modulator can be increased by using a loopfilter that only has a noise shaping function at these frequencies. This property of frequency asymmetry is

[2] Until now there is no commonly accepted way of combining IIP3, NF, G_T, ω_0 and P_{DC} in a figure-of-merit which only depends on technology variables (like e.g. the effective channel length). As a result, the performance comparison in this chapter does not take into account the evidently positive effect of a smaller technology on LNA performance; The performance metrics only compare the raw performance of the LNA's with respect to each other. The feature sizes of the technologies for the different circuits are listed in Table 6.4 and can be used for comparison.

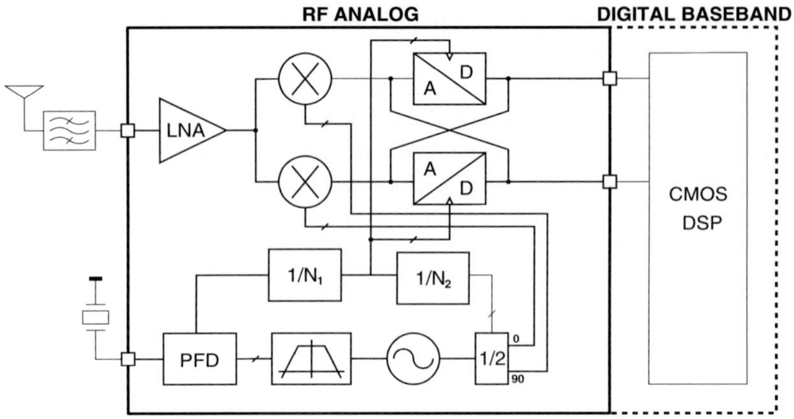

Figure 6.11: *High-level GPS receiver topology.*

exploited in this GPS receiver design [Ste02a]. A high-level schematic of the receiver is shown in Fig. 6.11

The receiver includes an ESD-protected LNA, a complex bandpass continuous time ADC and a fully integrated PLL with on-chip loop filter that generates both the LO-signals and the clock signals for the ADC. The only external components are a blocking filter and a reference crystal oscillator. No power hungry external LNA (30 mW) is required. A quadrature low-IF architecture is used to combine the advantages of a low IF architecture and a power efficient image rejection. Due to the wide dynamic range of the ADC, no VGA circuit is needed in the signal path. The high-level design considerations and techniques can be found in [Van03].

6.3.1.2 Low-Noise Amplifier

A more detailed circuit schematic of the LNA and downconverter is represented in Fig. 6.12. A single-ended LNA implementation is often preferred to a differential one in order to save power. An on-chip single-ended to differential balun placed directly after the LNA seems necessary but isn't appropriate because the junction capacitances and output impedances of the MOSFET may deteriorate the linearity performance. Instead, the single-ended to differential conversion is provided by the common mode feedback (CMFB) of the input opamp. At the duplicate node a replica of the LNA output impedance is placed to enhance the symmetry of the structure. The CMFB prevents this degradation of linearity. Moreover, removing the balun allowed to save area and power. The design of the LNA is discussed more thoroughly in Section 6.3.2

6.3.1.3 Quadrature, Direct Digital Downconversion

The 1.57 GHz RF-signal coming from the LNA is quadrature down-converted by mixers connected directly to the input of the A/D converter [Van02b]. The ADC is a continuous time $\Delta\Sigma$ A/D converter with a complex bandpass loop-filter. The output of the GPS receiver is a digital

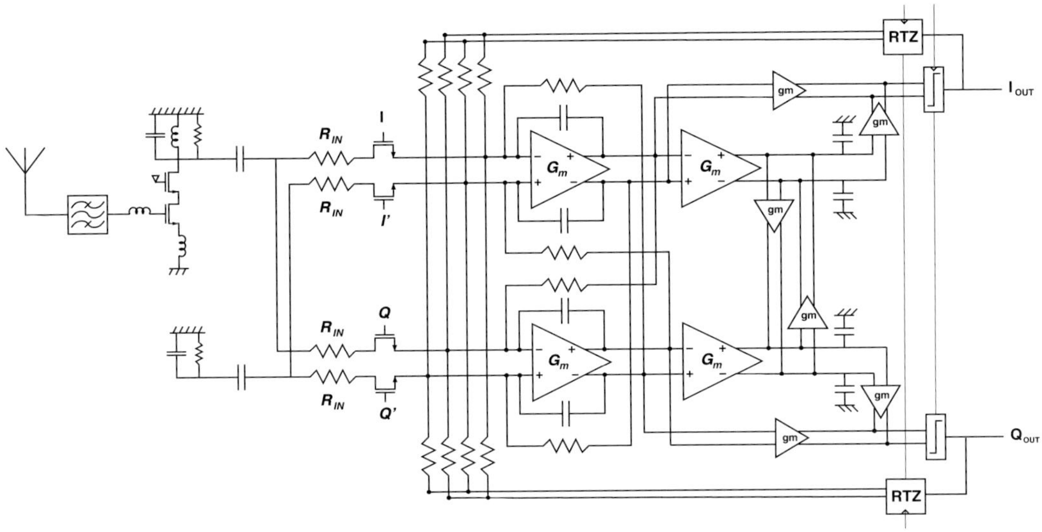

Figure 6.12: *Schematic of the LNA and downconversion mixer.*

I and Q bit-stream at a bit-rate of 128 MHz. Noise shaping Σ-Δ modulators are based on the oversampling technique and the loop filter characteristics to improve the resolution of the ADC's even with the use of a low resolution comparator. The complex bandpass ADC differs from its low-pass brother in the sense that it uses a complex bandpass filter, asymmetric around DC. With a passband around one positive (IF) frequency the quantization noise is shaped only for that band in contrary to a band-pass loop filter. In switched Σ-Δ A/D converters a complex anti alias filter is required in front of the ADC. This is no longer necessary for the continuous time implementation in this design.

6.3.1.4 PLL Frequency Synthesizer

The LO signal for the receiver is generated with a fully integrated fourth-order type-II PLL frequency synthesizer. The clock signals for the ADC are generated by the PLL as well. An LC-tank VCO as well as a 40 kHz LPF are integrated on-chip. The VCO operates at a frequency of 3.14 GHz. The quadrature signal is generated with a master-slave divide-by-2 block. The PLL is locked to a 16.37 MHz frequency reference through a divide-by-96 block and a phase frequency detector (PFD) without dead-zone. The reference frequency spurs are minimized by adding a reference branch in the charge pump core and by careful timing of the switch control signals. It is ensured that the charge pump current sources are always on. The current is alternatively flowing in the reference and the output branch of the charge pump. A virtual ground is provided after the charge pump by putting an OPAMP in the loop filter. This keeps the charge pump switches well in saturation and improves the symmetry between the Up- and the Down- side of the charge pump during locking. For stability reasons, a low frequency zero is inserted in the

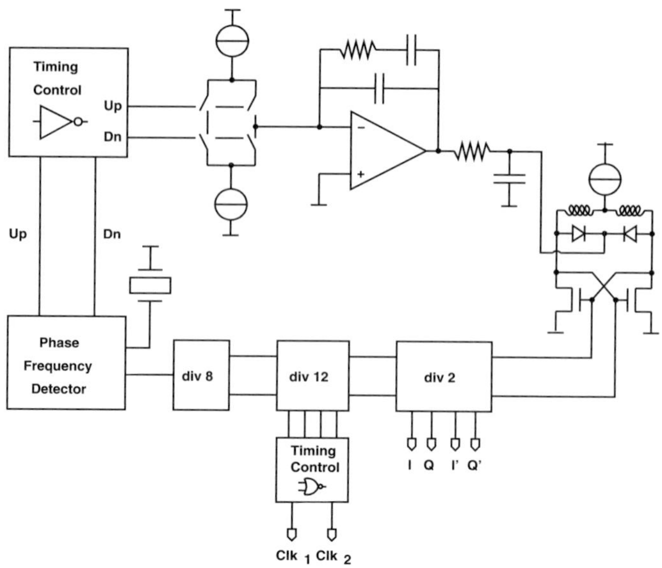

Figure 6.13: *Schematic of the PLL.*

loop filter. This comes at the cost of chip area since a huge capacitor of almost 2 nF needs to be integrated on chip.

6.3.2 The Low Noise Amplifier

The LNA for the L1 GPS receiver is shown in Fig. 6.14 [Ler02c]. The input ESD protection network consists of two diodes, D1 and D2, between the RF input and the power supply busses and of a stack of five diodes D3-D7 between V_{DD} and V_{SS}. The goal is to provide an explicit ESD discharge current path for all possible stress combinations. The use of diodes was based on the fact that they are very efficient and robust. Furthermore, their characteristics are fairly simple to model and simulate, allowing a reliable sizing of these devices. Since the LNA was designed for a fully integrated GPS receiver, where the output node of the LNA directly connects to the mixer input, no output ESD-protection is required.

All diodes have been implemented with several parallel squares. The area of the squares was chosen as small as possible while keeping more than one via for each square. The area of one diode square in this design is 9 μm^2. This yields a total area of 36 μm^2 for each input diode (one to V_{DD} and one to V_{SS}) corresponding to a total capacitance of 50 fF per diode . These values can also be found in Table 6.5.

The design of the diode string D3-D7 between the supply rails was based on the V_{DD} to V_{SS} leakage current specifications and the high current resistance requirements. Keeping in mind that both effects are influenced by the parasitic bipolar action of each diode, a series connection of five diodes is required. The size of the diodes has been chosen such that the on-resistance of

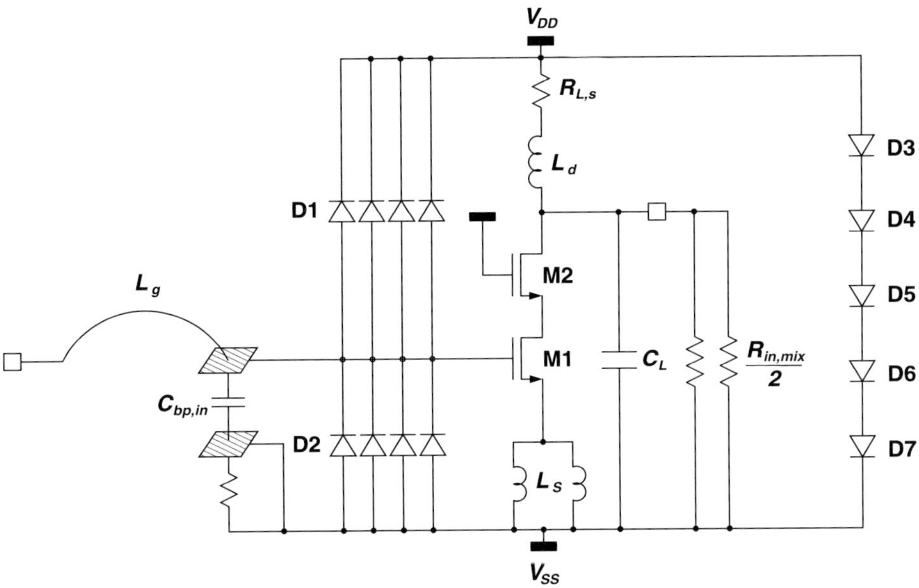

Figure 6.14: Simplified schematic of the LNA.

the clamp is lower than the on-resistance of the input diodes. Consequently, these diodes heave been sized five times larger than the input diodes. In this way, the supply clamp withstands higher ESD-stress than the input diodes ensuring that the latter is the limiting factor in the ESD-performance. The increased capacitance of the diode string does not pose a problem since it just adds to the on-chip decoupling capacitance and does not deteriorate the RF performance.

Fig. 6.15 plots the contour lines for the main performance parameters of the amplifier. The LNA is designed to directly drive the quadrature mixer input resistance. Therefore, it does no need to be matched to 50 Ω at the output. The main goal of the LNA is to have all possible input signal voltages amplified such that they fall within the dynamic range of the downconverter. It should do so with a minimum amount of noise addition and distortion. The value of the mixer degeneration resistance (R_{IN} in Fig. 6.12) largely determines the noise and distortion of the complete downconverter. Consequently this resistance more or less moves the dynamic range upwards or downwards. Enlarging the resistance will increase the noise but improve the linearity and therefore shift the dynamic range upward. Reducing the resistance will lower the noise but increase the distortion and shift the dynamic range downward. If the dynamic range is shifted upward, the voltage gain of the LNA will need to be increased and vice versa. For this chip the simulated voltage gain requirement was about 30 dB.

The behavior of the voltage gain A_v is plotted in Fig. 6.15(a). This behavior was obtained for a load resistance of 660 Ω excluding the input resistance of the mixer. This load resistance was obtained using an inductor of 10 nH with a series resistance of 15.5 Ω. The load resistance is higher than for the L2-band LNA discussed in Section 6.2. This is due to the fact that no

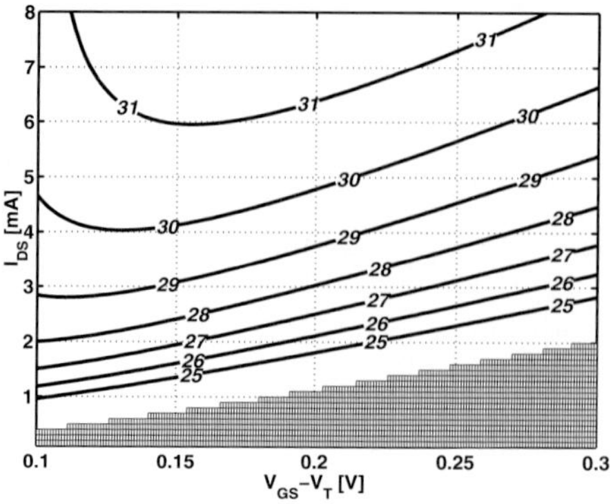

(a) A_v [dB].

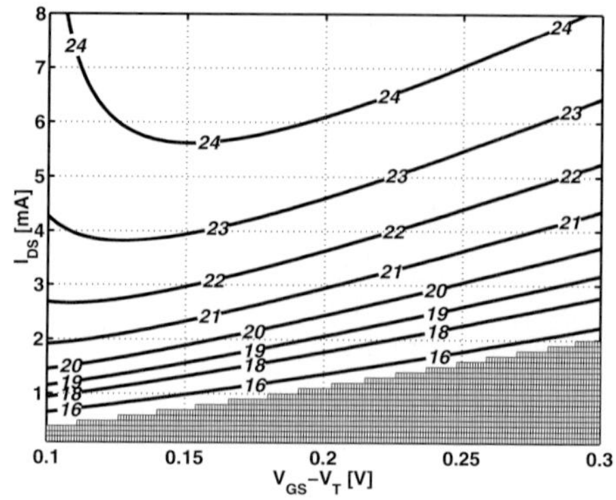

(b) G_T [dB].

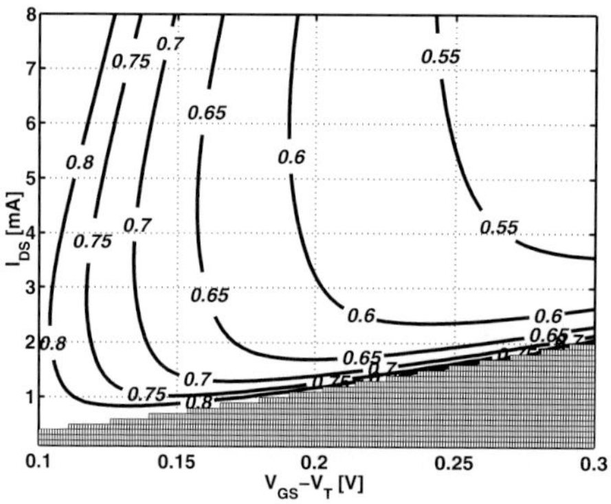

(c) NF [dB].

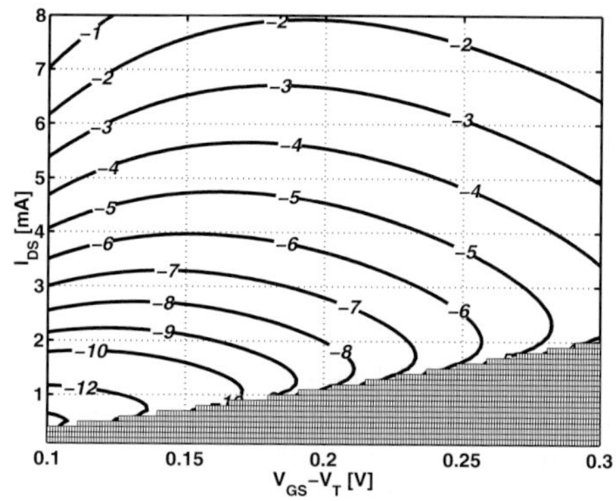

(d) IIP3 [dBm].

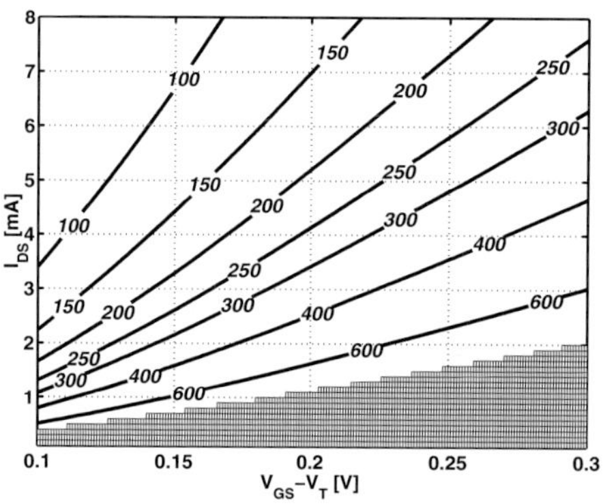

(e) $R_{S,opt}$ [Ω]

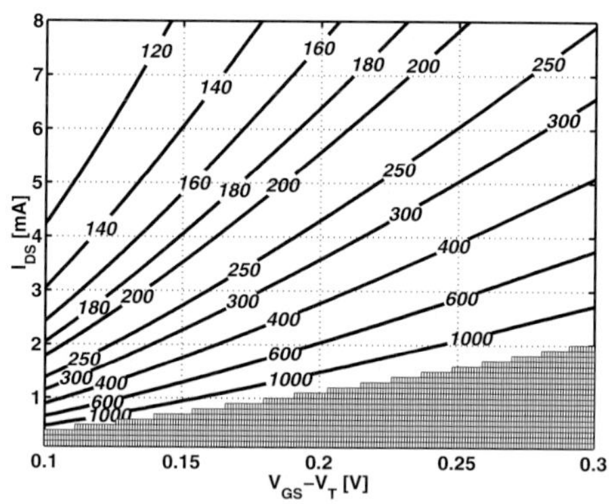

(f) $R_{S,eq}$ [Ω]

Figure 6.15: *Contour lines for the 1.57 GHz LNA.*

M1	I_{DS}	6 mA	M2	I_{DS}	6 mA
	$V_{GS} - V_T$	0.15 V		$V_{GS} - V_T$	0.2 V
	W/L	400/0.25		W/L	250/0.25
	g_m	60 mS		g_m	46 mS
	C_{gs}	290 fF		C_{gs}	190 fF
Gate inductance	L_g	17 nH	Load inductor	L_d	10 nH
Source inductance	L_s	1 nH		$R_{L,s}$	15.5 Ω
ESD diodes	A_D	2x36 μm^2		$R_{L,L}$	660 Ω
	P_D	2x48 μm	Input resistance mixer	$R_{in,mix}/2$	1000 Ω
	C_D	2x50 fF	Total load resistance	$R_{in,mix} \parallel R_{L,L}$	400 Ω
	R_D	~ 4 Ω	*RF bondpads*	C_{bp}	90 fF

Table 6.5: *Values for the main design parameters and components.*

capacitive divider is required here for matching the output to 50 Ω. As a consequence more capacitance is available and the self-resonance frequency of the inductor can be reduced. This option was used to decrease the series resistance maintaining the same inductance value.

At a current budget of 6 mA, the maximum voltage gain obtained in this way, is 31 dB at a $V_{GS} - V_T$ of 0.15 V. This corresponds to the chosen design point, listed in Table 6.5. The power gain is plotted in Fig. 6.15(b). It is about 24 dB a (0.15 V, 6 mA). The noise figure is shown in Fig. 6.15(c) and is about 0.7 dB. It could be increased for the same power budget by increasing the $V_{GS} - V_T$, i.e. reducing the width of M1. This is due to the relative dominance of the classical drain noise compared to the NQS noise. This dominance can also be deduced from a comparison between $R_{S,opt}$ and $R_{S,eq}$ in Fig. 6.15(d) and (e). The classical noise dominates whenever $R_{S,eq} > R_{S,opt}$. Going to larger overdrive voltages reduces the classical noise contribution due to the increase in ω_T. One can also see that $R_{S,eq}$ comes relatively closer to $R_{S,opt}$. The behavior of IIP3 is depicted in Fig. 6.15(e). The contribution of the drain-bulk capacitance of M2 is neglected since the bulk resistance is large enough owing to the specific layout of transistor M2. The IIP3 is about -3.5 dBm in the chosen design point. The main design parameters are listed in Table 6.5.

6.3.3 Results

This LNA was implemented in a standard 0.25 μm 4M1P CMOS process and occupies an area of 0.73 mm^2. A photograph of the IC is shown in Fig. 6.16. For the RF measurements, the LNA is glued on a ceramic substrate and is wire bonded to 50 Ω strip-lines. The substrate is then mounted in a Copper-Beryllium box, serving as a reference ground. The LNA is biased in two operating regimes drawing 4 mA and 6 mA from a 1.5 V supply.

The complete S-parameter set has been measured using an HP network analyzer. The power gain, S_{21}, is plotted in Fig. 6.17(a) for both operating regimes. The maximum power gain at 1.57 GHz is 16.5 dB and 15.5dB respectively. The input reflection of the circuit is shown in

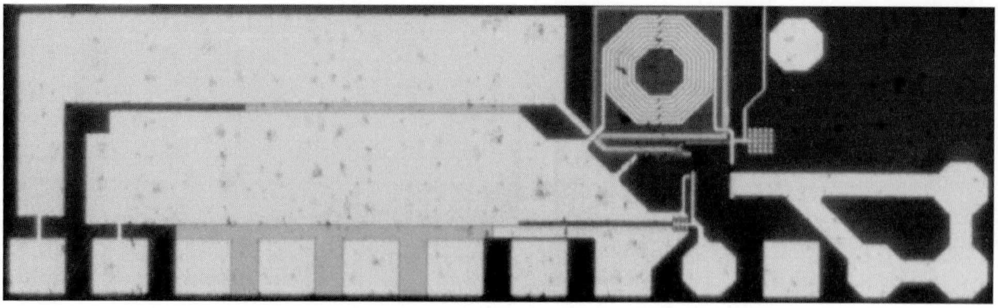

Figure 6.16: *Micrograph of the 1.57 GHz LNA.*

Supply Voltage	1.5 V	1.5 V
Current consumption	6 mA	4 mA
Power consumption	9 mW	6 mW
NF	1.3 dB	1.5 dB
S_{21}	16.5 dB	15.5 dB
S_{11}	$\sim$ -13 dB	$\sim$ -12 dB
S_{22}	—	—
S_{12}	$<$ -30 dB	$<$ -30 dB
IIP3	-5 dBm	-6 dBm

Table 6.6: *Compilation of the main RF measurement results for the LNA.*

Fig. 6.17(b). S_{11} is -13 dB and -12 dB respectively. The reverse isolation ($-S_{12}$) is measured to be larger than 30 dB throughout the entire range of the network analyzer (300 kHz - 3 GHz).

Fig. 6.17(c) and (d) depict the measured noise figure for both operating regimes. At 6 mA power consumption, the LNA has a NF of 1.3 dB at 1.57 GHz. In the 4 mA regime the NF is 1.5 dB. Fig. 6.17(e) shows the measured output power vs. input power for a two tone test in the 6 mA regime. The intercept point is at -5 dBm input power. At 4 mA the IIP3 is -6 dBm as depicted in Fig. 6.17(f). A summary of the RF-performance is given in Table 6.6.

Fig. 6.18 represents the measured Transmission Line Pulse (TLP) characteristics for the three most important stress combinations: INPUT to V_{DD}, INPUT to V_{SS} and V_{DD} to V_{SS}. The upper plot represents the high current IV characteristics for each stress condition. The lower plot shows the measured leakage current between the stressed pins. The jump in I_{leak} indicates the ESD failure threshold level has been reached. The results show that the TLP ESD robustness of both identical input diodes (Fig. 6.18(a)) is about 1.67 A. This corresponds to 2.5 kV Human Body Model (HBM) ESD stress. The diode resistance is in the order of 2 Ω. The ESD robustness of the D3-D7 diode stack is $\sim$ 3.25 A, corresponding to $\sim$ 4.9 kV HBM stress. The total resistance is also in the order of 2 Ω. The correlation of the TLP results to the HBM performance is for a first order evaluation only.

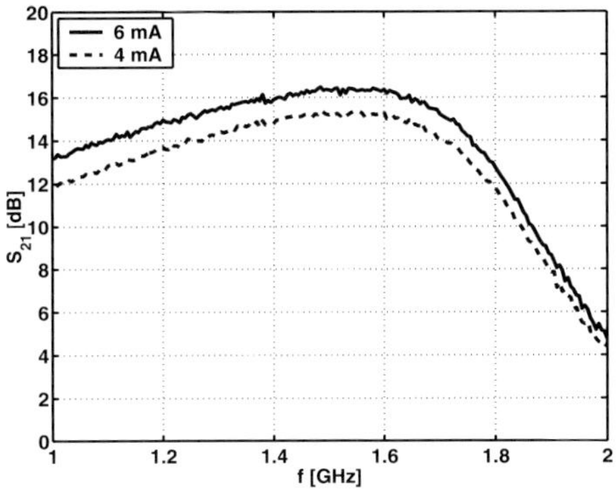

(a) Power gain [dB].

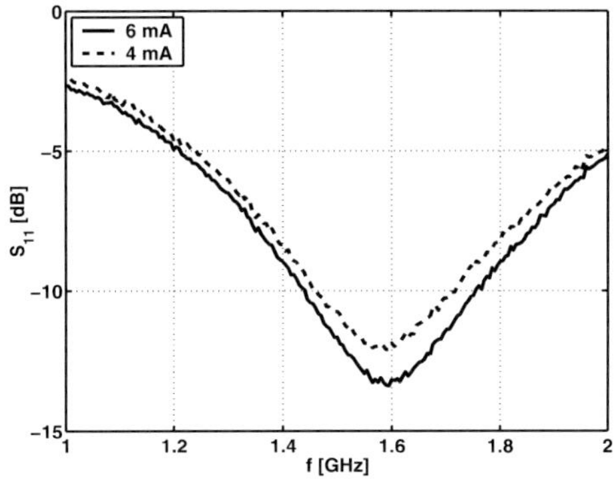

(b) Input reflection [dB].

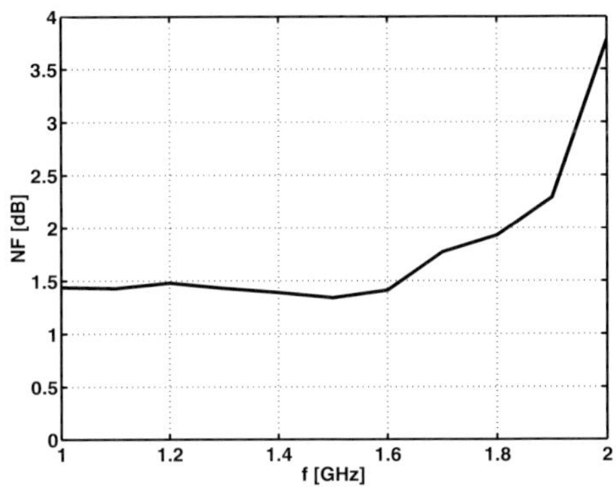

(c) Noise Figure @ 6 mA [dB].

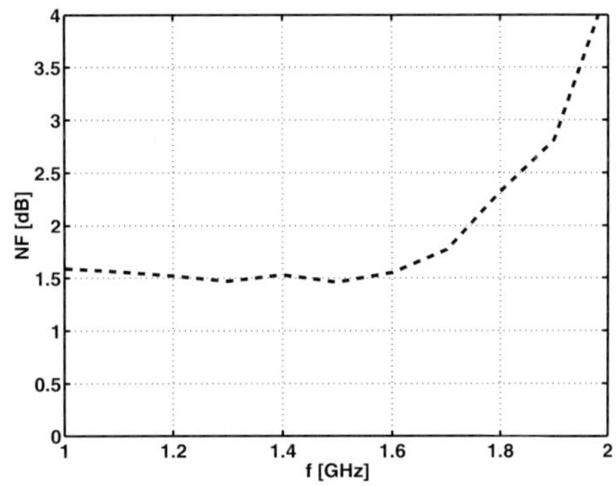

(d) Noise Figure @ 4 mA [dB].

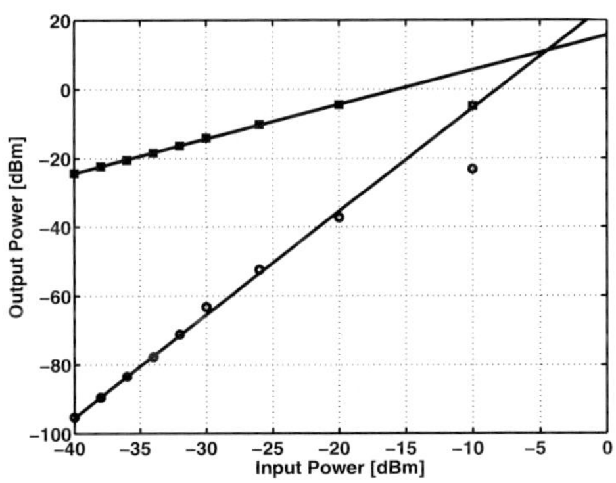

(e) Intermodulation distortion @ 6 mA.

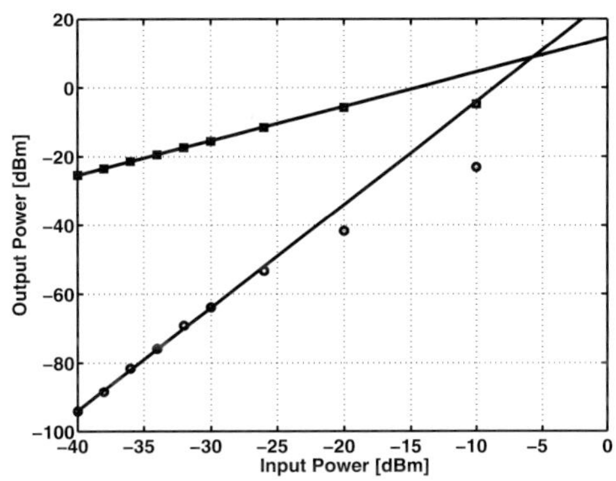

(f) Intermodulation distortion @ 4 mA.

Figure 6.17: *Measured RF performance for a current budget of 6 mA and 4 mA.*

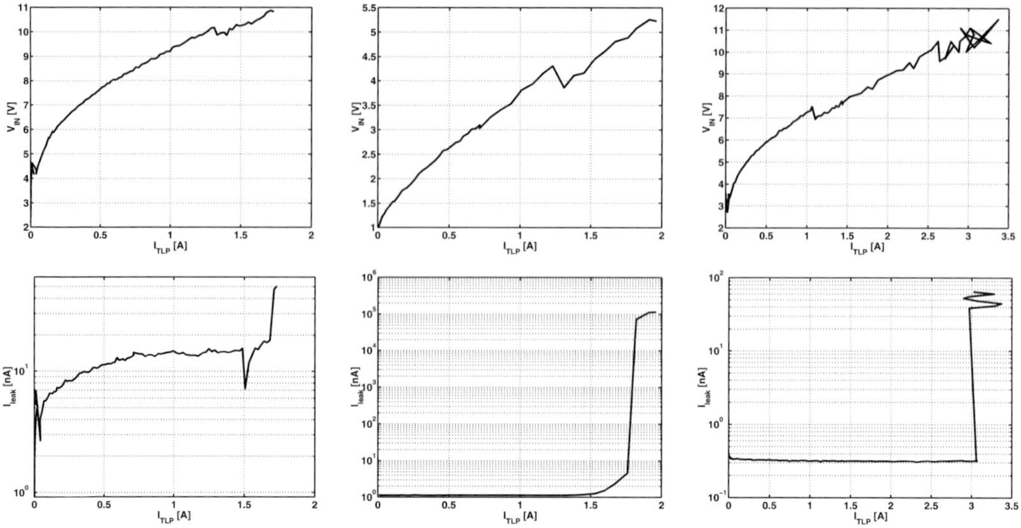

(a) IN- versus V_{SS} +. (b) IN+ versus V_{DD} +. (c) V_{DD} + versus V_{SS} +.

Figure 6.18: *Results for different TLP tests.*

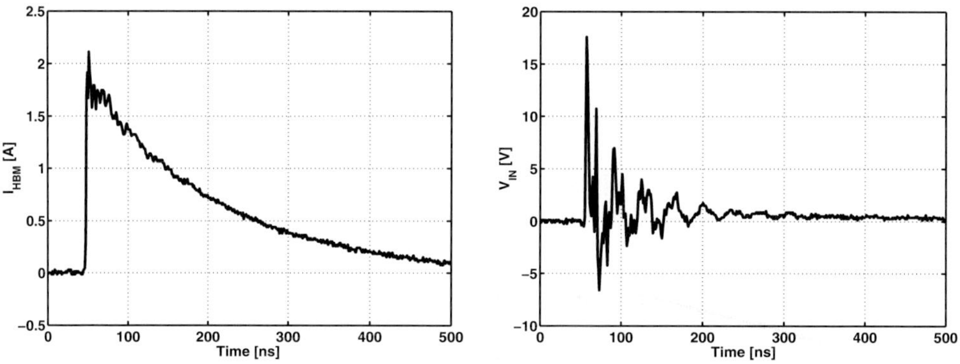

(a) HBM discharge current. (b) Corresponding input voltage.

Figure 6.19: *Typical HBM transient for an applied voltage of 3 kV.*

Stress combination	I_{TLP}	V_{HBM}
IN^{+} - ${V_{SS}}^{-}$	1.62 A	3.2 kV
IN^{+} - ${V_{DD}}^{-}$	1.67 A	3.2 kV
${V_{DD}}^{+}$ - ${V_{SS}}^{-}$	3.2 A	$>$ 4 kV
${V_{SS}}^{+}$ - ${V_{DD}}^{-}$	4.2 A	$>$ 4 kV

Table 6.7: *Summary of the main TLP and HBM results.*

LNA	Power consumption	8 mW
	NF	1.5 dB
	A_v	28 dB
	S_{11}	$\sim$ -12 dB
	IIP3	-6 dBm
PLL	Power consumption	17 mW (VCO: 10 mW)
	f_{ref}	16.37 MHz
	Phase noise	-115 dBc/Hz @ 600 kHz -138 dBc/Hz @ 3 MHz
	Locking range	10 % around 1.57 GHz
LO-buffer	Power consumption	2 mW
ADC	Power consumption	14.2 mW
	Frequency	0.3 - 1.6 GHz
	f_{CLK}	128 MHz
	f_{IF}	4 MHz
	BW	2 MHz
	OSR	32
	DR	62 dB
	Input noise level	-104 dBm
	IMRR	32 dB
Full receiver	Power consumption	37.2 mW
	Minimum Signal Level	-130 dBm
	Maximum Signal Level	-68 dBm
	Die area	16 mm^2

Table 6.8: Summary of the measurements results of the complete front-end.

Table 6.7 represents the ESD thresholds, achieved from the actual separate on wafer HBM testing. The worst case ESD stress combination for the circuit is when it is stressed between the input and ground nodes. In this case, the ESD current flows from the input pad through D1, the V_{DD} bus, the D3-D7 stack and the V_{SS} bus to the V_{SS} output pad. The overall ESD robustness is then determined by the lowest of the ESD thresholds of the different components in the current path: D1, D3-D7, the reverse breakdown of D2 and the ESD robustness of the input gate of the LNA. The measured ESD TLP threshold in this case was 1.62 A. Fig. 6.19 shows a typical HBM transient. The total discharge current is represented in Fig. 6.19(a). The corresponding voltage developed at the input is shown in Fig. 6.19(b). These graphs were obtained for the worst-case (first test in Table 6.7): IN+ vs. V_{SS}-. The results for the major stress combination are represented in Table 6.7. The other possible combinations can be represented by these results as well. The main measurement results for the complete receiver are listed in Table 6.8. A micrograph of the chip is shown in Fig. 6.20.

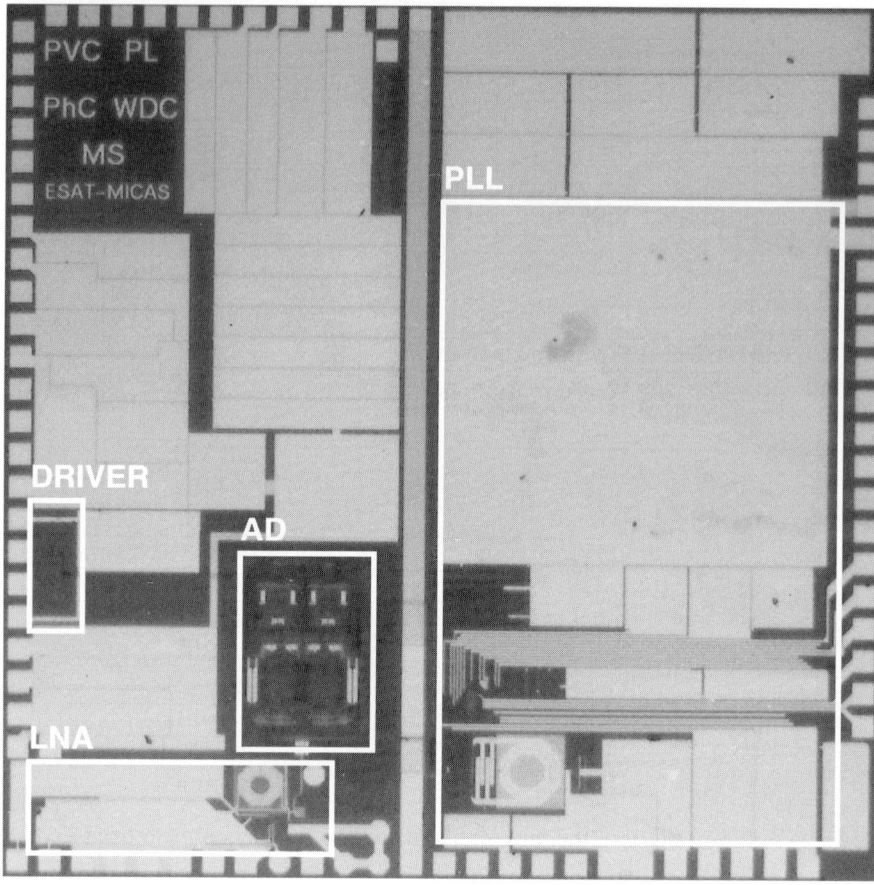

Figure 6.20: *Micrograph of the complete GPS receiver.*

6.3.4 Conclusion

This section has presented the design and measurement of a high performance 0.25 μm CMOS LNA for the L1 GPS band at 1.57GHz. The LNA features a 1.3 dB noise figure at 9 mW and a 1.5 dB noise figure at a mere 6 mW . The input ESD-protection is in the order of 3 kV HBM. This work has proven that, even in a standard submicron CMOS technology, a high RF-performance may be combined with a good level of ESD-protection satisfying the industrial specification of 2 kV HBM.

	IEEE 802.11a	HIPERLAN2
Frequency Band	5.15 - 5.35 GHz 5.725 - 5.825 GHz	5.15 - 5.35 GHz 5.47 - 5.725 GHz
Sensitivity	—	-70 dBm
Channel Bandwidth	—	24 MHz
Noise Figure	10 dB	18 dB @ SNR = 12 dB
Maximum Receive Power	-30 dBm	-25 dBm
ICP1	-26 dBm	-21 dBm
IIP3	-15 dBm	-10 dBm

Table 6.9: *Main specifications for the two 5 GHz Wireless LAN standards.*

6.4 A 5 GHz LNA with Inductive ESD-Protection Exceeding 3 kV HBM

6.4.1 5 GHz Wireless LAN

Two different standards are covered by the denominator 5 GHz wireless LAN. The first one is IEEE 802.11a. It occupies the frequency bands between 5.15 and 5.35 GHz and between 5.725 and 5.825 GHz. It allows bitrates up to 54 Mbps via an OFDM (Orthogonal Frequency Division Multiplexing) modulation scheme. Related standards are IEEE 802.11b which is already in a commercial phase. Operating at 2.45 GHz, it allows bitrates up to 11 Mbps via direct-sequence spread spectrum (DSSS). Another related standard is IEEE 802.11g which is a combination of both previous standards. It allows a bitrate up to 54 Mbps at 2.45 GHz.

The second standard is a European initiative, called HIPERLAN2 (High Performance LAN). It is very similar to IEEE 802.11a and defined in more or less the same frequency band: 5.15 - 5.35 GHz and 5.47 - 5.725 GHz. It also uses OFDM and achieves bitrates up to 54 Mbps. The main differences between the physical layers of both standards are listed in Table 6.9.

The required noise figure for both HIPERLAN2 and 802.11a receivers is a function of the data rate. Since it would be cumbersome to specify individual noise figures for each possible data rate, the specification for 802.11a instead simply recommends a noise figure of 10 dB, with a 5 dB implement margin, to accommodate the worst-case situation. For HIPERLAN2 the receiver sensitivity is defined for the given channel bandwidth of 24 MHz. The most stringent class C requirement specifies a sensitivity of -70 dBm. Assuming conservatively that the predetection SNR must exceed 12 dB, the overall receiver noise figure must be better than about 18 dB. As the IEEE 802.11a target is more demanding than that of HIPERLAN2, a 10 dB maximum noise figure should be the design goal for receiver front-ends complying with both standards.

The required IIP3 can be calculated from the maximum receive signal levels. Converting these signal levels into a precise IIP3 or 1 dB compression (ICP1) requirement is non trivial. However a simple but accurate approximation allows to define that the 1 dB compression point of the receiver should be about 4 dB above the maximum input signal. Based on this rule, we target a worst-case input-referred 1 dB compression point of -26 dBm and -21 dBm for

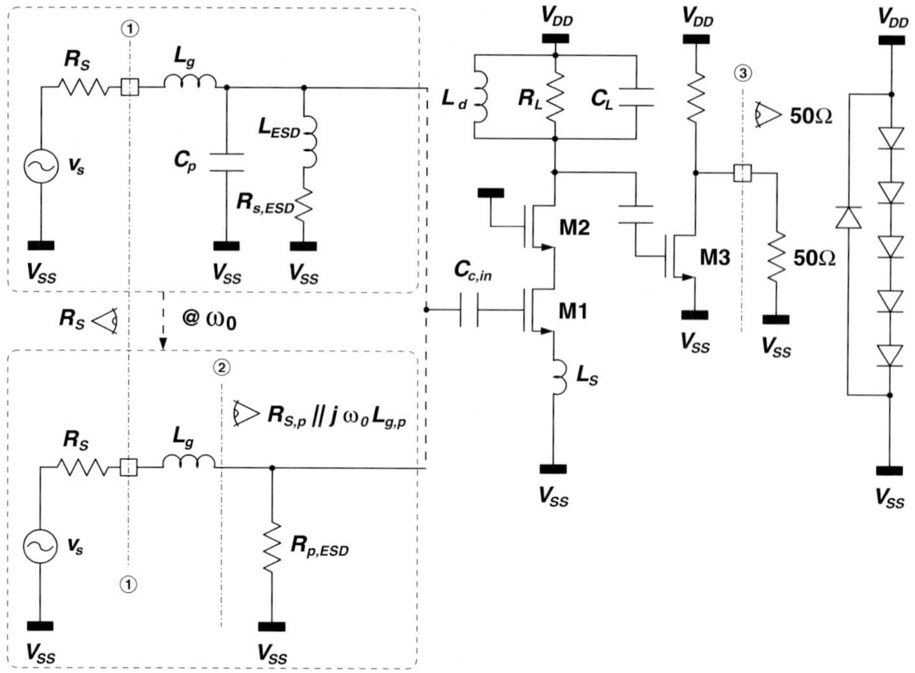

Figure 6.21: *Simplified schematic of the 5 GHz LNA with inductive input ESD-protection, supply clamp and 50 Ω output buffer.*

IEEE 802.11a and HIPERLAN2 respectively. The corresponding IIP3 values are -15 dBm and -10 dBm. Consequently, for large signals the HIPERLAN2 requirement is more stringent.

Notice that the total dynamic range (the IMFDR, defined by (2.73)) of the receiver should be larger for the IEEE 802.11a standard and the power levels are generally lower. For the HIPERLAN2 standard, the dynamic range can be less and the signal power levels are higher. A receiver aiming to comply with both standards should have an even wider dynamic range since it needs to be able to cope with the small signal levels from the IEEE 802.11a standard and the large signals from the HIPERLAN2 standard. Thus, the noise figure must be less than 10 dB and the IIP3 must be larger than -10 dBm.

6.4.2 Design

Fig. 6.21 shows the 5 GHz LNA employing an on-chip inductor, L_{ESD}, to provide ESD-protection. The output buffer provides an active 50 Ω termination. This active termination was preferred over a passive impedance transformation network for two reasons. Any passive transformation would further compress the bandwidth which is already rather small owing to the parallel resonance of the ESD inductor. Second, the buffer load resistance can be implemented by a MOS transistor in

K_n	V_{Tn}	Θ_n	Λ_n	α	α_{gd}	α_{gb}	$\alpha_{db} = \alpha_{sb}$ $V_{DB} = 0.5$ V	$\alpha_{db} = \alpha_{sb}$ $V_{DB} = 1.5$ V	γ	δ
[μA/V^2]	[V]	[V^{-1}]	[V^{-1}]	[ ]	[ ]	[ ]	[ ]	[ ]	[ ]	[ ]
491	0.5	6.55	0.15	0.87	0.27	0.12	0.70	0.55	2	4

L_{eff} [μm]	t_{ox} [nm]
0.147	4.2

Table 6.10: *Hand calculation parameters for the NMOS transistor in the 0.18 μm CMOS technology of UMC (extracted for $V_{GS} - V_T$ values between 0.1 and 0.3 V).*

the linear region which allows tuning after processing. The nominal output resistance has been set to 80 Ω instead of 50 Ω in order to be able to reduce the current consumption of the buffer for the same overall power gain.

The ESD inductor provides a bidirectional ESD path from input to ground. In order to protect the input for both positive and negative pulses versus both supply pins, a supply clamp has also been integrated. It has been implemented with a string of diodes, as discussed in Section 3.3.2.2. The number of diodes in the string depends on the operation voltage of the circuit determining the leakage and the high current resistance requirements. This resulted in a string of five diodes. The size of the diodes was calculated by setting the on-resistance of the diode clamp to ~ 3 Ω. All diodes consist of a parallel connection of several diode squares. Each square has a minimal area while still providing 4 contacts per square. The series resistance of these diodes is mainly due to the large n-well resistance. It is in the order of 100 Ω per square diode. The capacitance of this minimal diode is about 2.5 fF. This resulted in a total diode size of $\sim 130\ \mu\text{m}^2$ implemented as a parallel connection of 150 diodes of $\sim 0.88\ \mu\text{m}^2$. Five of those diodes in series represent a significant amount of chip area but they are located beneath the decoupling capacitors and no effective extra area is required. All other pins, including the output and biasing nodes, are protected with the same diodes, one to V_{SS}, one to V_{DD}. As a consequence, the IC is protected for any ESD event between any two pins. The main properties of the ESD diodes can be found in Table 6.11.

The design was done in a standard 0.18 μm 6M1P technology (Table 6.10). The primary design goal was to have an amplifier for the U-NII band (5.15-5.35 GHz). The center frequency was aimed at 5.25 GHz. The bandwidth should be large enough to cover the 200 MHz band with a minor gain offset and to cope with process variations. The frequency bands between 5.47 GHz and 5.725 GHz for HIPERLAN2 and between 5.725 GHz and 5.825 GHz for IEEE 802.11a are not considered in the design. Doing so would bring the total bandwidth up to 675 MHz. Taking into account the extra bandwidth necessary to cover process variations, the total bandwidth would need to be over 1 GHz. The corresponding Q-factor is 5 or less. This constraint would severely limit the gain of the LNA. Moreover since $R_{S,eq}$ would have to be increased also the noise figure would suffer or alternatively the power consumption would have to be increased. In this design the band of interest is limited to 5.15 - 5.35 GHz.

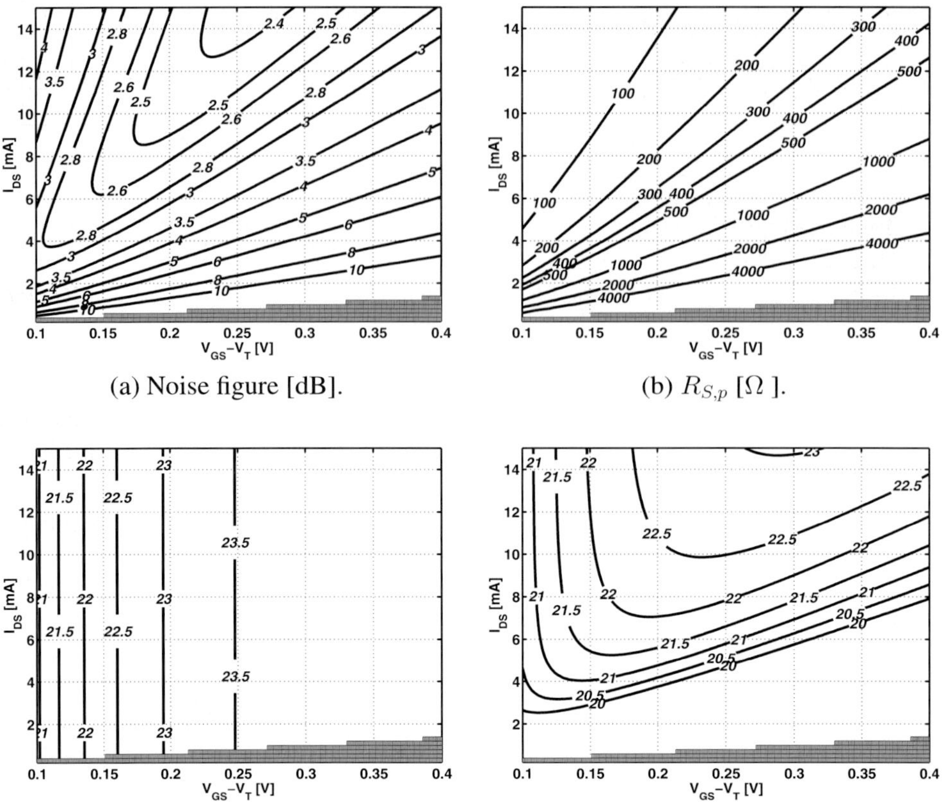

(a) Noise figure [dB]. (b) $R_{S,p}$ [Ω].

(c) LNA power gain without $R_{p,ESD}$ [dB]. (d) LNA power gain with $R_{p,ESD}$ [dB].

Figure 6.22: *Contours of noise figure,* $R_{S,p}$ *and* $R_{p,ESD}$ *for the 5 GHz LNA with ESD inductor.*

The ESD inductor was designed to take an ESD current of at least 2 Amps during a relatively short period. The voltage drop over the inductor should remain sufficiently low in order to safeguard the nearby gate oxide. The DC breakdown voltage in a 0.18 μm technology is about 3.5 V. For ESD, which is a short term event, the maximum *'safe'* voltage is at least double. Hence, the voltage drop over the inductor needs to remain below 7 V which corresponds to 3.5 Ω for 2 A. Note that this resistance is more or less the DC resistance. At 5 GHz, the series resistance of the inductor will be larger. This is mainly due to the Skin-effect but also partly to eddy-current losses in the ground shield and substrate. The inductor is implemented as an octagonal coil and not as a rectangular coil. Therefore all corners have an angle of 135°instead of 90°. This is beneficial for the large ESD currents through the coil which are then spread more evenly over the coil width and avoid hot spots.

The addition of the buffer has to be incorporated in the behavioral model of the LNA. The buffer has been designed such that the stand-alone voltage gain, $A_{v,buf}$ is 0 dB. This means

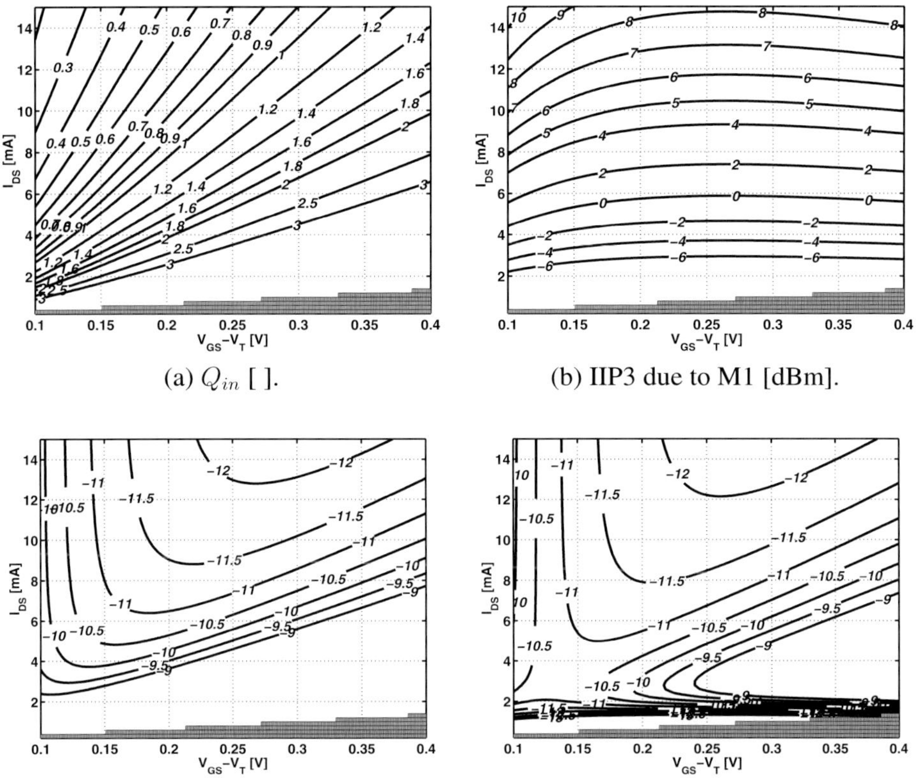

(a) Q_{in} [ ].

(b) IIP3 due to M1 [dBm].

(c) IIP3 due to M3 [dBm].

(d) Total IIP3 [dBm].

Figure 6.23: *Contour plots of the input quality factor and the IIP3 contributions for the 5 GHz LNA with ESD inductor.*

the voltage gain is -4 dB when the buffer is connected to the external 50 Ω load. The parameter values and components of the output buffer can be found in Table 6.11. Note that even though the voltage gain is negative, the extra power gain, given by $10\log(g_{m,buf}^2 R_L R_{L,buf})$ is still positive (+5 dB) as will be discussed later. The buffer also introduces a minor extra contribution to the noise factor stemming from the channel noise of M3:

$$F_{buf} - 1 = 4\left(\frac{\omega_0}{\omega_T}\right)^2 \frac{\gamma R_{S,eq}}{g_{m,buf} R_L^2}, \tag{6.3}$$

where $g_{m,buf}$ is the transconductance of buffer transistor M3. The noise of the buffer load resistance can be neglected. Fig. 6.22(a) shows the total noise figure contours of the LNA in the design space of M1. In these calculations the excess noise factor γ is set to two as indicated in Table 6.10. The equivalent load resistance connected to the drain of M2 is chosen about 300 Ω. This value is rather low in order not to jeopardize the stability of the circuit. The load resistance

M1	I_{DS}	8 mA	M2	I_{DS}	8 mA
	$V_{GS}-V_T$	0.165 V		$V_{GS}-V_T$	0.22 V
	W/L	240/0.18		W/L	160/0.18
	g_m	70 mS		g_m	55 mS
	C_{gs}	320 fF		C_{gs}	220 fF
M3	I_{DS}	4 mA	Buffer	$R_{L,buf}$	80 Ω
	$V_{GS}-V_T$	0.4 V		$A_{v,buf}$	0 dB
	W/L	36/0.18	Inductors	L_d	1.2 nH
	g_m	13 mS		$R_{L,s}$	4.5 Ω
	C_{gs}	47 fF		$R_{L,L}$	320 Ω
ESD diodes	$A_{D,\square}$	0.88 μm^2		L_g	1.9 nH
	$P_{D,\square}$	3.76 μm		L_s	0.26 nH
	$C_{D,\square}$	~ 2.5 fF		L_{ESD}	2 nH
	$R_{D,\square}$	~ 100 Ω		$R_{s,ESD}$ @ DC	3.2 Ω
	squares per diode	150		$R_{s,ESD}$ @ 5 GHz	4 Ω
	diodes in supply clamp	5		$R_{p,ESD}$ @ 5 GHz	1 kΩ
	resistance supply clamp	~ 3 Ω			

Table 6.11: *Values for the main design parameters and components.*

was implemented by an on-chip inductor in parallel with a linear pMOS transistor. The pMOS transistor was added to allow gain-tuning in case of stability problems. The patterned regions indicate where no input match can be obtained as described by (4.5) and (4.28). The inductance of the ESD inductor has been chosen such that it tunes out the parasitic input capacitance completely:

$$L_{ESD} = \frac{1}{\omega_0^2 C_p}, \tag{6.4}$$

where C_p incorporates the bonding pad capacitance, the wiring capacitance, the gate-drain capacitance of M1 and the parasitic capacitance of the ESD-inductor itself. In other words the effective parasitic input capacitance remaining is zero and $R_{S,eq} = R_S = 50\ \Omega$. This choice minimizes the noise contribution from the ESD inductor as shown by (5.22). The total parasitic capacitance is in the order of 500 fF which yields an inductor of 2 nH. As a consequence,

$$R_{p,ESD} = \frac{\omega_0^2 L_{ESD}^2}{R_{s,ESD}} \approx 1\ \text{k}\Omega, \tag{6.5}$$

for $\alpha_{ind} = 2\ \Omega/\text{nH}$. The amplifier was designed for a current consumption of 8 mA at a power supply of 1.5 V. The optimum $V_{GS} - V_T$ for this current is 0.165 V.

The power gain is somewhat different for the circuit in Fig. 6.21 compared to the G_T equation in Table 5.4. This is due to the presence of the output buffer:

$$G_T = \frac{g_{m,buf}^2 R_L^2 R_{L,buf}}{4R_S} \left(\frac{\omega_T}{\omega_0}\right)^2 \frac{1}{(1 + M\alpha_{gd})^2}, \tag{6.6}$$

where $R_{L,buf} = 80\ \Omega$ is the output resistance of the output buffer. The contour lines of the gain are plotted in Fig. 6.22(d). According to (6.6) the power gain should only be function of $V_{GS} - V_T$, not of the current. Hence the contours should be vertical lines. However, this is only true for any design point where the equivalent parallel resistance of the ESD inductor is large compared to the equivalent parallel source resistance. If this is not the case, then the power gain is approximated by

$$G_T \approx \frac{g_{m,buf}^2 R_L^2 R_{L,buf}}{R_S} \left(\frac{\omega_T}{\omega_0}\right)^2 \frac{1}{(1 + M\alpha_{gd})^2} \left(\frac{R_{p,ESD}}{2R_{p,ESD} + R_{S,p}}\right)^2, \qquad (6.7)$$

where it is assumed that the input match is perfect for $R_{p,ESD} = \infty$. Hence, the gain will be unaffected if $2R_{p,ESD} \gg R_{S,p}$. Otherwise a significant amount of input signal power is lost in the ESD-inductor and the power gain decreases. This is illustrated also in Fig. 6.22. Fig. 6.22(c) shows the ideal gain for $R_{p,ESD} = \infty$. Fig. 6.22(d) shows the gain for $R_{p,ESD} = 1\ \text{k}\Omega$. At the chosen design point (0.165 V, 8 mA), the gain is only minorly reduced from 22.5 dB to 22 dB. Fig. 6.22(b) shows that $R_{S,p}$ is low for large width devices corresponding to a small gate inductor. It decreases towards the upper left. This is also the region where the gain contours will start to go vertical in Fig. 6.22(d). For $R_{p,ESD} = 1\ \text{k}\Omega$ the boundary is more or less at $R_{S,p} = 200\ \Omega$ and close to the chosen design point .

The passive gain, Q_{in}, provided by the tuned input stage and calculated by (4.57) is rather low, owing to the relatively high frequency of operation. It is plotted in Fig. 6.23(a) and is about 0.7 at the chosen design point (0.165 V, 8 mA). This low value is beneficial for the bandwidth of the circuit. The parallel resonance introduced by the ESD-inductor will be the main bandwidth limiter. This is due to the very large quality factor of the resonance. In this design it is given by $\frac{\omega_0 L_{ESD}}{R_{s,ESD}} \approx 15$.

The IIP3 for transistor M1, plotted in Fig. 6.23(b) is high and in the vicinity of 3 dBm at the design point. This is mainly due to the low Q_{in} which limits the voltage across the gate-source capacitance of M1. In this design, the non-linearity of M3 is dominant. Transistor M3 has to cope with much larger signal levels. The corresponding input referred IP3 is shown in Fig. 6.23(c). Only at very low current levels, the M1 non-linearity is dominant. This is due to the low gain of the first stage. The total IIP3 is plotted in Fig. 6.23(d) and is only marginally distinct from Fig. 6.23(c). The main design parameters and component values can be found in Table 6.11.

6.4.3 Results

The LNA has been implemented in a standard 0.18 μm 6M1P CMOS technology. A micrograph of the chip is shown in Fig. 6.24. The ESD tests were performed directly by probing the wafer. Both Transmission Line Pulse (TLP) and Human Body Model (HBM) measurements have been done with different stress polarities between the different pins. The main test results are indicated in Table 6.12. The HBM protection level of all pin combinations exceeds 3 kV. TLP results even show that the ESD-devices can handle currents up to more than 3 Amps.

For the RF-measurements, the LNA was mounted on a ceramic substrate using a flip-chip technique to contact the 50 Ω strip lines. The substrate was placed in a copper-beryllium box

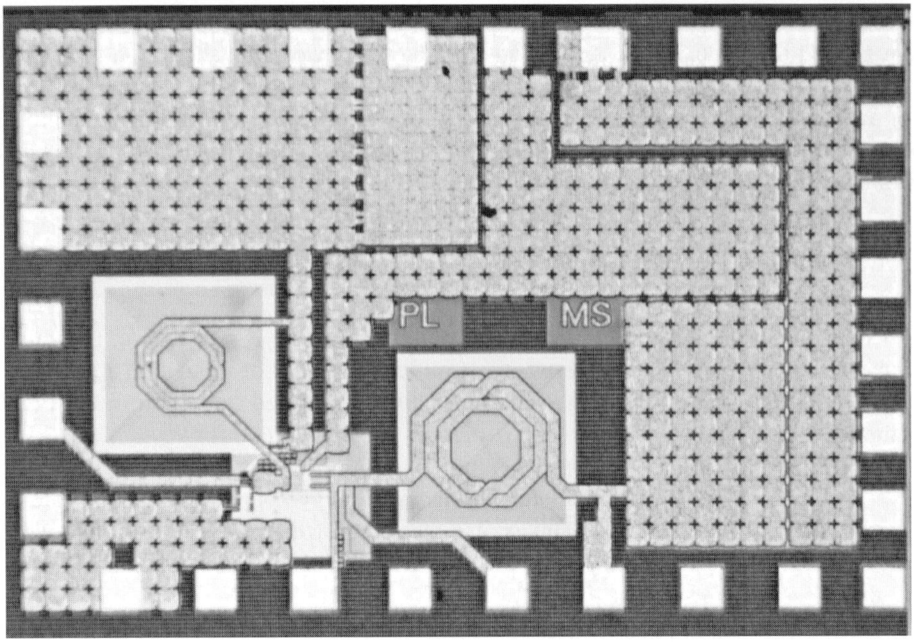

Figure 6.24: *Micrograph of the 5 GHz LNA.*

which serves as a reference ground. SMA connectors are used to connect the RF-pins to the outside world.

The LNA itself draws 8 mA from a 1.5V supply. The 50 Ω buffer consumes an extra 4 mA. A complete S-parameter set was recorded for frequencies up to 8 GHz. Input and output reflection coefficients are given in Fig. 6.25(a) and (b). They are both approximately -20 dB at 5 GHz. The power gain of the LNA is indicated in Fig. 6.25(c) and reaches a maximum of 20 dB. This ensures a very good rejection of the mixer noise. The reverse gain plotted in Fig. 6.25(d) is lower than -35dB from 3.5 to 6.5 GHz and lower than -30 dB throughout the measurement range. The measured noise figure is shown in Fig. 6.25(e). This measurement was performed without any de-embedding of substrate parasitics. The calibration was done at the level of the SMA connectors. The minimum noise figure is 3.5 dB and is achieved at 5 GHz. The linearity of the circuit was measured with a two-tone test. The input IP3 was found to be -9 dBm. This is only just sufficient for the wireless LAN system specifications in Section 6.9. However, both the calculations in Section 6.4.2 and more detailed simulations have shown that the IIP3 is limited by the non-linearity of M3 in the output buffer. If the LNA is integrated with the down-conversion mixer, the buffer will not be required and the IIP3 is expected to be in the order of 5 dBm. A summary of the RF-performance is given in Table 6.13.

The measurements still show a significant offset with the calculations in Section 6.9. This is mainly due to the simplification of the design environment. These crude calculations were checked with numerical simulations. The center frequency for this design was aimed at 5.25 GHz.

Stress combination	I_{TLP}	V_{HBM}
IN^+ - V_{SS}^-	$>$ 3 A	$>$ 3 kV
IN^+ - V_{DD}^-	$>$ 3 A	$>$ 3 kV
V_{DD}^+ - V_{SS}^-	$>$ 4 A	$>$ 3 kV
V_{SS}^+ - V_{DD}^-	$>$ 4 A	$>$ 3 kV

Table 6.12: *Main ESD results.*

Supply Voltage	1.5 V
Current consumption	8 mA + 2 mA buffer
Power consumption	15 mW
NF	3.5 dB
S_{21}	20 dB
S_{11}	$\sim$ -20 dB
S_{22}	$\sim$ -20 dB
S_{12}	$<$ -30 dB
IIP3	-9 dBm

Table 6.13: *Experimental results at 5 GHz.*

The actually measured center frequency was 5 GHz. This 5 % offset probably owes to an underestimation of the parasitic capacitances, possibly due to process variations. The difference in simulated and measured gain was 2 dB. This could be largely explained by a small increase of the series resistance of the load inductor and/or the ESD inductor.

The IIP3 value from the measurements is -9 dBm compared to -11 dBm seen in the contour plots. This difference of 2 dB is equal to the difference in gain. This supports the presumption that the IIP3 is limited by the second stage and that the 2 dB gain is lost in the first stage. The 2 dB gain reduction decreases the signal levels at the gate of M3 with 2 dB and hence increases the IIP3 accordingly. The value of IIP3 in simulation behaved rather erratically. This is probably due to the constructive or destructive interference of several non-linearities. A minor change in one of the design parameters could change the IIP3 with up to 5 dB.

The measured noise figure is 3.5 dB, while it was just over 2.5 dB in the calculations and 2.7 dB in simulations. Several possible reasons can be found to explain this difference.

- There has been no de-embedding of substrate parasitics. The resistance in the strip line connecting the input of the LNA was not taken into account. This can be in the order of a few Ω.

- The resistance in the poly gate fingers of M1 could be larger than the simulated value.

- The transconductance of the LNA can be different.

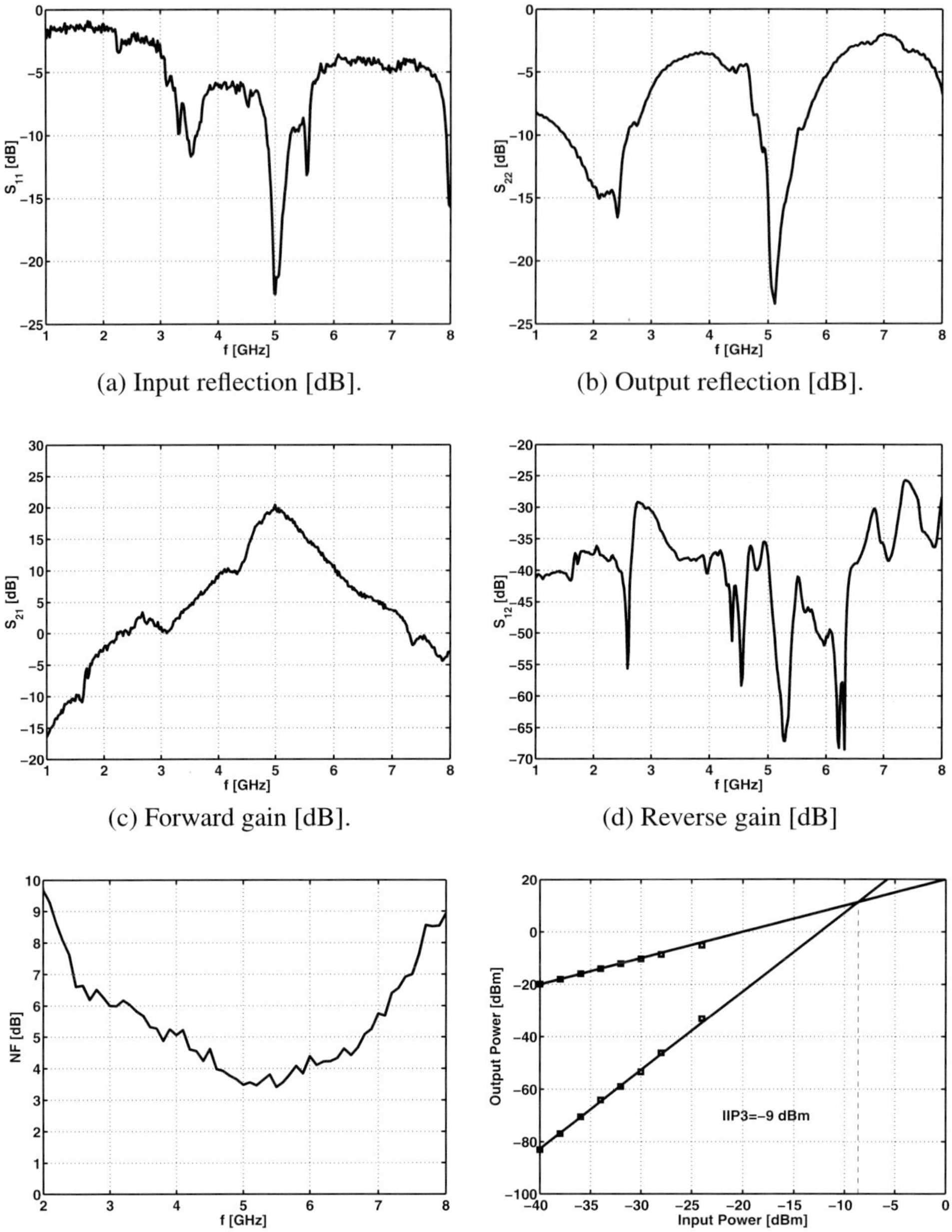

(a) Input reflection [dB].

(b) Output reflection [dB].

(c) Forward gain [dB].

(d) Reverse gain [dB]

(e) Noise figure [dB].

(f) Two tone output power [dBm].

Figure 6.25: *Measured S-parameters, noise figure and IIP3 for the 5 GHz LNA.*

- The excess noise factor can be higher than the estimated value of 2. Especially for small technologies, the electric field in the channel can be quite large and cause noisy hot electrons.
- The NQS gate resistance can be higher than estimated. Indeed, κ is equal to 5 for long channel transistors, but could be lower for short-channel transistors.
- There could be noise leakage through the substrate coupling back into the input of the amplifier or steering the back-gate of M1.

6.4.4 Conclusion

ESD-protection parasitics at the input of a low-noise amplifier are often detrimental for the RF-performance. Especially as the frequency of operation increases. This problem has been tackled by using an on-chip inductor for ESD-protection. This inductor is able to drain any ESD-current up to more than three Amps. At the RF-frequency, the inductor is designed such that it tunes out the harmful parasitic capacitance at the input. The 5 GHz LNA, matched at both input and output, has a power gain of 20 dB and a noise figure of 3.5 dB. The IIP3 is limited by the 50 Ω output buffer and measured -9 dBm. The power consumption is 15 mW including the output buffer.

6.5 Conclusion

Several integrated low-noise amplifiers have been discussed. The first amplifier has been presented in Section 6.2. It operates at 1.23 GHz and is matched to 50 Ω on-chip at both input and output. The 0.8 dB noise figure is the lowest NF published to date for a CMOS LNA. It is achieved with a power consumption of 9 mW. The corresponding power gain is 20 dB.

Section 6.3 describes the design and measurement of a low-noise amplifier for the GPS L1 band. It has been integrated with a complete GPS receiver front-end. The amplifier features a bidirectional input ESD-protection of 3 kV HBM. The noise figure measures 1.3 dB for a power consumption of 9 mW and 1.5 dB for 6 mW.

A 5 GHz LNA for wireless LAN applications has been presented in Section 6.4. It features an integrated inductor at the input for ESD-protection. The noise figure of the amplifier is 3.5 dB. The corresponding power gain is 20 dB. The ESD-protection level has a minimum of 3 kV HBM. This chip is to the authors' knowledge, the first competitive CMOS LNA at 5 GHz with on-chip ESD-protection exceeding the industrial 2 kV HBM standard.

Chapter 7

Conclusions

The continuous expansion of the telecommunication market has increased the demand for low cost transceivers. This evolution presents the main driving force behind recent research on integrated CMOS solutions. The research in this work focusses on RF low-noise amplifiers for wireless communication. One of the remaining challenges in this domain is to provide good RF-performance with respect to noise, gain and linearity with a circuit that is fully protected against ESD. The primary goal was to investigate the inherent potential of CMOS technologies to achieve the required performance at different frequencies under these ESD constraints. Two LNA circuits have been designed for a portable GPS receiver. The GPS system is a very demanding application owing to the extremely low signal levels which should be received. This makes it an ideal demonstrator. A second RF LNA was designed to demonstrate the potential of CMOS for 5 GHz applications, even under ESD constraints.

- A very thorough study of the common-source LNA with inductive source degeneration has exposed the importance of several parasitic components. The impact of all relevant parasitic components has been incorporated in a compact but complete behavioral model of the amplifier. This model comprises several design equations directly translating the design parameter choices to the true performance measures: noise figure, power gain and IIP3.

- This analysis has revealed the severe impact of the parasitic input capacitance, C_p, on both noise figure and gain. It has been shown that C_p even places an upper bound on the frequency at which an input match can be obtained without an additional matching network. The input ESD protection devices further increase this parasitic capacitance, reduce the performance and lower the cut-off frequency. Other solutions are required at frequencies above roughly 2 GHz.

- Two topologies have been proposed that succeed in generating both an input-match on-chip and protecting the circuit against ESD for much higher operation frequencies. The first topology is based on the Π-type matching network where the capacitors are replaced by the parasitic capacitors of the ESD-devices. This matching network creates an additional degree of freedom which allows to tailor the equivalent source resistance into whatever yields the best performance for the given application. The second topology uses a parallel inductor to provide ESD-protection. The low-frequency ESD-pulses are bypassed to ground while at RF the inductor is designed to tune out any or all parasitic input capacitance depending on the required performance.

- Based on the foregoing analysis, several low-noise amplifiers have been integrated in mainstream CMOS technologies. The first amplifier operates at 1.27 GHz and is matched to 50 Ω on-chip at both input and output. The 0.8 dB noise figure is the lowest NF published to date for a CMOS LNA. It is achieved with a power consumption of 9 mW. The corresponding power gain is 20 dB. This IC proves that even in a standard CMOS technology LNA's can be realized that have performance comparable to commercial GaAs implementations.

  A second amplifier has been integrated within a complete 1.57 GHz L1 GPS receiver front-end. The amplifier features a bidirectional input ESD-protection of 3 kV HBM. The noise figure measures 1.3 dB for a power consumption of 9 mW and 1.5 dB for 6 mW. This LNA has shown that a high RF performance can be combined with a good level of ESD-immunity surpassing the industrial 2 kV specification.

  The third design that was discussed regards a 5 GHz LNA for wireless LAN applications. It features an integrated inductor at the input providing ESD-protection. The noise figure of the amplifier measures 3.5 dB. The corresponding power gain is 20 dB. The ESD-protection level has a minimum of 3 kV HBM. This chip is to the authors' knowledge, the first competitive CMOS LNA at 5 GHz with on-chip ESD-protection exceeding the industrial 2 kV HBM standard.

Appendix A

Fundamentals of Two-Port Noise Theory

Any noisy two-port can be replaced with a noiseless two-port with two input noise sources depending only on the noisy two-port itself. The equivalence of both is valid for any source impedance. This equivalent two-port is shown in Fig. A.1. The two noise sources are the noise current source, $\overline{i_n^2}$ between the positive and negative input and the noise voltage source, $\overline{e_n^2}$ in series with either input. The source impedance is represented by G_s and jB_s which are the resistive and reactive component of the source admittance. The noise of the source impedance is represented by $\overline{i_s^2}$.

Since the input noise current i_n may be partly correlated with the input noise voltage, e_n, the input noise current is split up according to

$$i_n = i_c + i_u, \tag{A.1}$$

where i_c and i_u represent the correlated and uncorrelated part of the input noise current respectively. The correlation allows to rewrite i_c as

$$i_c = Y_c e_n, \tag{A.2}$$

where

$$Y_c = G_c + jB_c \tag{A.3}$$

is the complex correlation admittance. Reconfiguring the different noise sources into their Norton equivalent allows to add them together through superposition:

$$i_{n,tot} = i_s + i_c + i_u + Y_s e_n = i_s + i_u + (Y_c + Y_s)e_n. \tag{A.4}$$

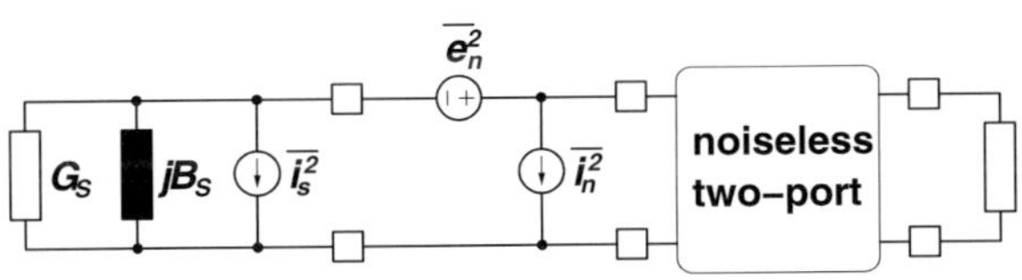

Figure A.1: *Equivalent noise model of a two-port and its signal source.*

This yields three noise current sources which are mutually uncorrelated. The total average squared noise current, $\overline{i_{n,tot}^2}$ is found as

$$\overline{i_{n,tot}^2} = \overline{i_s^2} + \overline{i_u^2} + |Y_c + Y_s|\overline{e_n^2}. \tag{A.5}$$

where the cross averages are zero. The noise factor can be calculated by dividing this total squared noise current by the squared noise current of the source:

$$F = 1 + \frac{\overline{i_u^2} + |Y_c + Y_s|^2\overline{e_n^2}}{\overline{i_s^2}}. \tag{A.6}$$

This can be rewritten as

$$F = 1 + \frac{G_u + |Y_c + Y_s|^2 R_n}{G_s}, \tag{A.7}$$

where

$$R_n \triangleq \frac{\overline{e_n^2}}{4kT\Delta f} \qquad G_u \triangleq \frac{\overline{i_u^2}}{4kT\Delta f} \qquad G_s \triangleq \frac{\overline{i_s^2}}{4kT\Delta f}. \tag{A.8}$$

G_c, B_c, G_u and R_n are four noise parameters completely describing the noise behavior of the original two-port.

The value of Y_s that optimizes the noise factor F is denoted by $Y_{opt} = G_{opt} + jB_{opt}$ —the optimum source admittance— and is found by differentiation:

$$G_{opt} = \sqrt{\frac{G_u}{R_n} + G_c^2} \tag{A.9}$$

$$B_{opt} = -B_c. \tag{A.10}$$

The minimum noise factor can be calculated by substituting (A.9) in (A.7)

$$F_{min} = 1 + 2R_n(G_{opt} + G_c). \tag{A.11}$$

Equation (A.7) can now be rewritten as

$$F = F_{min} + \frac{R_n}{G_s}|Y_s - Y_{opt}|^2. \tag{A.12}$$

F_{min}, B_{opt}, G_{opt} and R_n are four equivalent noise parameters again completely describing the noise behavior of the original two-port.

Index

Bibliography

[Abi95] A. A. Abidi, "Direct Conversion Radio Transceivers for Digital Communications", *IEEE Journal of Solid-State Circuits*, vol. 30, no. 12, pp. 1399–1410, July 1995.

[Aie03] G. R. Aiello, "Challenges for Ultra-wideband (UWB) CMOS Integration", in *Proceedings of the IEEE Radio Frequency Integrated Circuits Symposium*, pp. 497–500, June 2003.

[Ame99] A. Amerasekera et al., "Analysis of snapback behaviour on the ESD capability of sub-20 μm NMOS", in *Proceedings of the IEEE International Reliability Physics Symposium*, 1999.

[Aok02] I. Aoki, S. Kee, D. Rutledge, and A. Hajimiri, "Fully-Integrated CMOS Power Amplifier Design Using the Distributed Active Transformer Architecture", *IEEE Journal of Solid-State Circuits*, vol. 37, no. 3, pp. 371–383, Mar. 2002.

[Bru00] F. Bruccoleri, E. A. M. Klumperink, and B. Nauta, "Generating All 2-Transistor Circuits Leads to New Wide-Band CMOS LNAs", in *Proceedings of the IEEE European Solid-State Circuits Conference*, Stockholm, Sweden, Sept. 2000.

[Bru02] F. Bruccoleri, E. A. M. Klumperink, and B. Nauta, "Noise Canceling in Wideband CMOS LNAs", in *Proceedings IEEE International Solid-State Circuits Conference*, pp. 406–407, San Francisco, USA, Feb. 2002.

[Cas02] D. J. Cassan and J. R. Long, "A 1V 0.9dB NF Low Noise Amplifier for 5-6GHz WLAN in 0.18 m CMOS", in *Proceedings Custom Integrated Circuits Conference*, Orlando, Florida, USA, May 2002.

[Cas03] D. J. Cassan and J. R. Long, "A 1-V Transformer Feedback Low-Noise Amplifier for 5-GHz Wireless LAN in 0.18 μm CMOS", *IEEE Journal of Solid-State Circuits*, vol. 38, no. 3, pp. 427–435, Mar. 2003.

[Cha91] Z. Y. Chang and W. M. C. Sansen, *Low-Noise Wide-Band Amplifiers in Bipolar and CMOS Technologies*, Kluwer Academic Publishers, Boston, 1991.

[Cha02] C.-Y. Cha and S.-G. Lee, "A 5.2GHz LNA in 0.35 μm CMOS Utilizing Inter-stage Series Resonance and Optimizing the Substrate Resistance", in *Proceedings of the*

IEEE European Solid-State Circuits Conference, pp. 339–342, Firenze, Italy, Sept. 2002.

[Cha03] C.-Y. Cha and S.-G. Lee, “A 5.2GHz LNA in 0.35 μm CMOS Utilizing Inter-stage Series Resonance and Optimizing the Substrate Resistance”, *IEEE Journal of Solid-State Circuits*, vol. 38, no. 4, pp. 669–672, Apr. 2003.

[Col99] J. T. Colvin, S. S. Bhatia, and K. O. Kenneth, “Effects of Substrate Resistances on LNA Performance and a Bondpad Structure for Reducing the Effects in a Silicon Bipolar Technology”, *IEEE Journal of Solid-State Circuits*, vol. 34, no. 9, pp. 1339–1344, Sept. 1999.

[Cro96] J. Crols, P. Kinget, J. Craninckx, and M. Steyaert, “An Analytical Model of Planar Inductors on Lowly Doped Silicon Substrates for High Frequency Analog Design up to 3 GHz”, in *Proceedings of the IEEE Symposium on VLSI Circuits*, pp. 28–29, Hawaii, USA, June 1996.

[Cro98] J. Crols and M. Steyaert, “Low-IF topologies for High-Performance Analog Front-Ends for Fully Integrated Receivers”, *IEEE Transactions on Circuits And Systems II*, vol. 45, no. 3, pp. 269–282, Mar. 1998.

[Dab98] S. Dabral and T. J. Maloney, *Basic ESD and I/O Design*, Wiley Interscience, New York, 1998.

[Din01] Y. Ding and R. Harjani, “A +18 dBm IIP3 LNA in 0.35 μm CMOS”, in *Proceedings IEEE International Solid-State Circuits Conference*, pp. 162–163, San Francisco, USA, Feb. 2001.

[Duv95] C. Duvvury and A. Amerasekera, “Advanced CMOS Protection Device Trigger Mechanisms During CDM”, in *Proceedings EOS/ESD Symposium*, page 162, 1995.

[Edw92] M. L. Edwards and J. H. Sinsky, “A New Criterion for Linear 2-Port Stability Using a Single Geometrically Derived Parameter”, *IEEE Transactions on Microwave Theory and Techniques*, Dec. 1992.

[Eld01] *Mentor Graphics Eldo User Manual, Release 2001.2*, September 2001.

[Enz02] C. Enz, “MOS Transistor Modeling for RF Integrated Circuit Design”, in *Proceedings of the Fifth International Conference on Modeling and Simulation of Microsystems, Workshop on Compact Modeling*, San Juan, Puerto Rico, U.S.A., Apr. 2002.

[Flo99] B. A. Floyd, J. Mehta, and C. Gamero, “A 900 MHz 0.8 μm CMOS Low Noise Amplifier with 1.2 dB Noise Figure”, in *Proceedings Custom Integrated Circuits Conference*, pp. 661–664, San Diego, USA, May 1999.

[Fra03] L. M. Franca-Neto, B. A. Bloechel, and K. Soumyanath, "17GHz and 24GHz LNA Designs based on Extended-S-parameter with Microstrip-on-Die in 0.18μm Logic CMOS Technology", in *Proceedings of the IEEE European Solid-State Circuits Conference*, Estoril, Portugal, Sept. 2003.

[Gal03] S. Galal and B. Razavi, "Broadband ESD Protection Circuits in CMOS Technology", *IEEE Journal of Solid-State Circuits*, vol. 38, no. 12, pp. 2334–2340, Dec. 2003.

[GPS96] *Global Positioning System: Theory and Applications, Volume I (i.e. Progress in Aeronautics and Astronautics, Vol. 163)*, edited by B. W. Parkinson, J. J. Spilker Jr., P. Axelrad and P. Enge, American Institute of Aeronautics and Astronautics, Washington, 1996.

[Gra00a] G. Gramegna, A. Magazzu, C. Sclafani, and M. Paparo, "Ultra-Wide Dynamic Range 1.75 dB Noise-Figure, 900MHz CMOS LNA", in *Proceedings IEEE International Solid-State Circuits Conference*, pp. 380–381, San Francisco, USA, Feb. 2000.

[Gra00b] G. Gramegna, A. Magazzu, C. Sclafani, M. Paparo, and P. G. Erratico, "A 9 mW, 900 MHz CMOS LNA with 1.05 dB Noise Figure", in *Proceedings of the IEEE European Solid-State Circuits Conference*, pp. 155–158, Stockholm, Sweden, Sept. 2000.

[Gra01] G. Gramegna, M. Paparo, P. G. Erratico, and P. De Vita, "A sub-1-dB NF 2.3-kV ESD-Protected 900-MHz LNA", *IEEE Journal of Solid-State Circuits*, vol. 36, no. 7, pp. 1010–1017, July 2001.

[Gua02] X. Guan and A. Hajimiri, "A 24 GHz CMOS Front-end", in *Proceedings of the IEEE European Solid-State Circuits Conference*, pp. 155–158, Firenze, Italy, Sept. 2002.

[Hau60] H. A. Haus et al., "Representation of noise in linear two-ports", *Proc. IRE*, vol. 48, pp. 69–74, Jan. 160.

[Hay98] G. Hayashi, H. Kimura, H. Simomura, and A. Matsuzawa, "A 9 mW 900 MHz CMOS LNA with Mesh Arrayed MOSFETs", in *Proceedings of the IEEE Symposium on VLSI Circuits*, pp. 84–85, Honolulu, Hawaï, June 1998.

[HSp01] *Avant! Star-Hspice Manual, Release 2001.2*, June 2001.

[Hua98a] Q. Huang, P. Orsatti, and F. Piazza, "Broadband, 0.25 μm CMOS LNAs with Sub-2dB NF for GSM Applications", in *Proceedings Custom Integrated Circuits Conference*, pp. 67–70, Santa Clara, USA, May 1998.

[Hua98b] Q. Huang, F. Piazza, P. Orsatti, and T. Ohguro, "The Impact of Scaling Down to Deep Submicron on CMOS RF Circuits", *IEEE Journal of Solid-State Circuits*, vol. 33, no. 7, pp. 1023–1036, July 1998.

[inb93] *ESD Sensitivity Tesiting: Human Body Model (HBM)–Component Level*, chapter S5.1, ESD Association Standard, 1993.

[inb94] *ESD Sensitivity Tesiting: Machine Model (MM)–Component Level*, chapter S5.2, ESD Association Standard, 1994.

[inb99] *ESD Sensitivity Tesiting: Charged Device Model (CDM)–Component Level*, chapter S5.3, ESD Association Standard, 1999.

[Ito02] C. Ito, K. Banerjee, and R. W. Dutton, "Analysis and Design of Distributed ESD Protection Circuits for High-Speed Mixed-Signal and RF ICs", *IEEE Transactions on Electron Devices*, vol. 49, no. 8, pp. 1444–1454, Aug. 2002.

[Jan97] J. Janssens, M. Steyaert, and H.Miyakawa, "A 2.7 Volt CMOS Broadband Low Noise Amplifier", in *Proceedings of the IEEE Symposium on VLSI Circuits*, pp. 87–88, Kyoto, June 1997.

[Jan98] J. Janssens, J. Crols, and M. Steyaert, "A 10 mW Inductorless, Broadband Low Noise Amplifier for 900 MHz Wireless Communications", in *Proceedings Custom Integrated Circuits Conference*, pp. 75–78, Santa Clara, USA, May 1998.

[Jan99a] J. Janssens and M. Steyaert, "Design of Broadband Low-Noise Amplifiers in Deep-Submicron CMOS Technologies", in *Analog Circuit Design. 1 Volt electronics, mixed-mode systems, low-noise and RF power amplifiers for telecommunication.* edited by J. Huijsing, R. van de Plassche and W. Sansen, Kluwer Academic Publishers, Boston, pp. 317-335, 1999.

[Jan99b] J. Janssens and M. Steyaert, "MOS noise performance under impedance matching constraints", *IEE Electronics Letters*, vol. 35, no. 5, pp. 1278–1280, July 1999.

[Jan99c] J. Janssens and M. Steyaert, "Optimum MOS Power Matching by Exploiting Non-Quasi Static Effect", *IEE Electronics Letters*, vol. 35, no. 8, pp. 672–673, Apr. 1999.

[Jan01] J. Janssens, *Deep Submicron Cellular Receiver Front-Ends*, PhD thesis, K.U.Leuven, Belgium, July 2001.

[Jin86] R. P. Jindal, "Hot electron effects on channel thermal noise in fine line NMOS field effect transistors", *IEEE Transactions on Electron Devices*, vol. 32, no. 9, pp. 1395–1397, Sept. 1986.

[Kam94] M. Kamon, M. J. Tsuk, and J. K. White, "FASTHENRY: A Multipole-Acclerated 3-D Inductance Extraction Program", *IEEE Transactions on Microwave Theory and Techniques*, Sept. 1994.

[Kar96] A. N. Karanicolas, "A 2.7V 900MHz CMOS LNA and Mixer", in *Proceedings IEEE International Solid-State Circuits Conference*, pp. 50–51, San Francisco, USA, Feb. 1996.

[Kis99] S. V. Kishore, G. Chang, G. Asmanis, C. Hull, and F. Stubbe, "Substrate-Induced High-Frequency Noise in Deep Submicron MOSFETs for RF Applications", in *Proceedings Custom Integrated Circuits Conference*, pp. 365–368, San Diego, USA, May 1999.

[Lak94] K. R. Laker and W. M. C. Sansen, *Design of Analog Integrated Circuits and Systems*, MCGraw-Hill, New York, 1994.

[Lee98] T. H. Lee, *The Design of CMOS Radio Frequency Integrated Circuits*, Cambridge University Press, 1998.

[Lee00] D. Leenaerts and P. de Vreede, "Influence of substrate noise on RF performance", in *Proceedings of the IEEE European Solid-State Circuits Conference*, pp. 300–303, Stockholm, Sweden, Sept. 2000.

[Lee03] D. Leenaerts and R. Velghe, "Bond pad and esd protection structure for 0.25 μm/ 0.18 μm RF-CMOS", in *Proceedings of the IEEE European Solid-State Circuits Conference*, Estoril, Portugal, Sept. 2003.

[Ler01a] P. Leroux, J. Janssens, and M. Steyaert, "A 0.8 dB NF ESD-protected 9 mW CMOS LNA", in *Proceedings IEEE International Solid-State Circuits Conference*, pp. 410–411, San Francisco, Feb. 2001.

[Ler01b] P. Leroux and M. Steyaert, "A high performance 5.2GHz LNA with an on-chip inductor to provide ESD-protection", *IEE Electronics Letters*, vol. 37, no. 5, pp. 467–469, Mar. 2001.

[Ler02a] P. Leroux, J. Janssens, and M. Steyaert, "A 0.8dB NF ESD-protected 9mW CMOS LNA Operating at 1.23 GHz", *IEEE Journal of Solid-State Circuits*, vol. 37, no. 6, pp. 760–765, June 2002.

[Ler02b] P. Leroux and M. Steyaert, "A New ESD Protection Topology for High Frequency CMOS Low Noise Amplifiers", in *Proceedings of the IEEE International Symposium on Electromagnetic Compatibility*, pp. 129–133, Sorrento, Italy, Sept. 2002.

[Ler02c] P. Leroux, V. Vassilev, M. Steyaert, and G. Groeseneken, "A 1.3dB NF CMOS LNA for GPS with 3kV HBM ESD-protection", in *Proceedings of the IEEE European Solid-State Circuits Conference*, pp. 335–338, Firenze, Italy, Sept. 2002.

[Ler02d] P. Leroux, V. Vassilev, M. Steyaert, and H. Maes, "A 6 mW, 1.5 dB NF CMOS LNA for GPS with 3 kV HBM ESD Protection", in *Proceedings EOS/ESD Symposium*, pp. 18–25, Charlotte, NC, USA, Oct. 2002.

[Ler03a] P. Leroux and M. Steyaert, "RF-ESD Co-Design for High Performance CMOS LNAs", in *Proceedings of the Workshop on Advances in Analogue Circuit Design*, Graz, Austria, Apr. 2003.

[Ler03b] P. Leroux and M. Steyaert, "RF-ESD Co-Design for High-Performance CMOS LNAs", in *Analog Circuit Design. Fractional-N Synthesizers, Design for Robustness, Line and Bus Drivers.* edited by A. H. M. van Roermund, M Steyaert and J. Huijsing, Kluwer Academic Publishers, Boston, 2003.

[Ler03c] P. Leroux, V. Vassilev, M. Steyaert, and H. Maes, "High ESD Performance, Low Power CMOS LNA for GPS Applications", *IEEE Journal of Electrostatics*, vol. 59, no. 3-4, pp. 179–192, Apr. 2003.

[Lon00] J. R. Long, "Monolithic Transformers for Silicon RF IC Design", *IEEE Journal of Solid-State Circuits*, vol. 35, no. 9, pp. 1368–1382, Sept. 2000.

[Mag03] A. Magnusson and H. Olsson, "A 1.8-V wide-band cmos lna for multiband multistandard front-end receiver", in *Proceedings of the IEEE European Solid-State Circuits Conference*, Estoril, Portugal, Sept. 2003.

[Mal95] T.J. Maloney and S. Dabral, "Novel Clamp Circuits for IC Power Supply Protection", in *Proceedings EOS/ESD Symposium*, 1995.

[Man98] T. Manku, "Microwave CMOS - Devices and Circuits", in *Proceedings Custom Integrated Circuits Conference*, pp. 59–66, Santa Clara, USA, May 1998.

[Man99] T. Manku, M. Obrecht, and Y. Lin, "RF Simulations ans Physics of the Channel Noise Parameters within MOS Transistors", in *Proceedings Custom Integrated Circuits Conference*, pp. 365–368, San Diego, USA, May 1999.

[Ors99] P. Orsatti, F. Piazza, Q. Huang, and T. Morimoto, "A 20mA-Receive, 55mA-Transmit GSM Transceiver in 0.25μm CMOS", in *Proceedings IEEE International Solid-State Circuits Conference*, San Francisco, USA, Feb. 1999.

[Pia98] F. Piazza, P. Orsatti, and Q. Huang, "A 0.25μm CMOS Transceiver Front-End for GSM", in *Proceedings Custom Integrated Circuits Conference*, pp. 413–416, Santa Clara, USA, May 1998.

[Pol92] T. Polgreen and A. Chatterjee, "Improving the ESD Failure Threshold of Silicided n-MOS Output Transistors by Ensuring Uniform Current Flow", *IEEE Trans. RD.*, vol. 39, no. 2, pp. 379, 1992.

[Rad01] M. Radhakrishnan et al., "ESD reliability issues in RF CMOS circuits", in *Proceedings of the IEEE International Workshop on the Physics of Semiconductor Devices*, Dec. 2001.

[Raz94] B. Razavi, T. H. Yan, and K. F. Lee, "Impact of distributed gate resistance on the performance of MOS devices", *IEEE Transactions on Circuits and Systems I*, vol. 41, no. 11, pp. 750–754, Nov. 1994.

[Raz97] B. Razavi, "Design Considerations for Direct Conversion Receivers", *IEEE Transactions on Circuits And Systems II*, vol. 44, no. 6, pp. 428–453, June 1997.

[Ric00] C. Richier et al., "Investigation on Different ESD protection strategies devoted to 3.3V RF applications in a 0.18 μm CMOS process", in *Proceedings EOS/ESD Symposium*, pp. 251–259, Sept. 2000.

[Rof96] A. Rofougaran, J. Y.-C. Chang, M. Rofougaran, and A. A. Abidi, "A 1 GHz CMOS RF Front-End IC for a Direct-Conversion Wireless Receiver", *IEEE Journal of Solid-State Circuits*, vol. 31, no. 7, pp. 880–889, July 1996.

[Rof98a] A. Rofougaran et al., "A Single-Chip 900MHz Spread-Spectrum Wireless Transceiver in 1μm CMOS - part I: Architecture and Transmitter Design", *IEEE Journal of Solid-State Circuits*, vol. 33, no. 4, pp. 515–534, Apr. 1998.

[Rof98b] A. Rofougaran et al., "A Single-Chip 900MHz Spread-Spectrum Wireless Transceiver in 1μm CMOS - part II: Receiver Design", *IEEE Journal of Solid-State Circuits*, vol. 33, no. 4, pp. 535–547, Apr. 1998.

[Sam99] H. Samavati, H. Rategh, and T. Lee, "A 12.4mW CMOS front-end for a 5GHz wireless-LAN receiver", in *Proceedings of the IEEE Symposium on VLSI Circuits*, pp. 87–90, June 1999.

[San99] W. Sansen, "Distortion in elementary transistor circuits", *IEEE Transactions on Circuits And Systems*, vol. 46, pp. 315–325, Mar. 1999.

[Sev94] J. Sevenhans et al., "An Analog Radio Front-End Chip Set for a 1.9GHz Mobile Radio Telephone Application", in *Proceedings IEEE International Solid-State Circuits Conference*, pp. 44–45, San Francisco, USA, Feb. 1994.

[Sha97a] D. K. Shaeffer and T. H. Lee, "A 1.5 V, 1.5 GHz, CMOS Low Noise Amplifier", *IEEE Journal of Solid-State Circuits*, vol. 32, no. 5, pp. 745–759, May 1997.

[Sha97b] A. R. Shahani, D. K. Shaeffer, and T. H. Lee, "A 12-mW Wide Dynamic Range CMOS Front-End for a Portable GPS Receiver", *IEEE Journal of Solid-State Circuits*, vol. 32, no. 12, pp. 2061–2070, Dec. 1997.

[Ste96] M. Steyaert, M. Borremans, J. Craninckx, J. Crols, J. Janssens, and P. Kinget, "RF CMOS Design: Some Untold Pitfalls", in *Analog Circuit Design. MOST RF circuits, sigma-delta converters and translinear circuits.* edited by W. Sansen, J. Huijsing and R. van de Plassche, Kluwer Academic Publishers, Boston, pp. 63-88, 1996.

[Ste98] M. Steyaert, M. Borremans, J. Janssens, B. De Muer, and N. Itoh et al., "A single-chip CMOS transceiver for DCS-1800 wireless communications", in *Proceedings IEEE International Solid-State Circuits Conference*, pp. 48–49, San Francisco, USA, Feb. 1998.

[Ste99] M. Steyaert, M. Borremans, J. Janssens, and B. De Muer, *"RF Communication Circuits",The VLSI Handbook*, chapter 65, 1999.

[Ste00a] M. Steyaert, J. Janssens, B. De Muer, M. Borremans, and N. Itoh, "A 2V CMOS cellular transceiver front-end", in *Proceedings IEEE International Solid-State Circuits Conference*, pp. 142–143, San Francisco, Feb. 2000.

[Ste00b] M. Steyaert, J. Janssens, B. De Muer, M. Borremans, and N. Itoh, "A 2 Volt CMOS Cellular Transceiver Front-End", *IEEE Journal of Solid-State Circuits*, vol. 35, no. 12, pp. 1895–1907, Dec. 2000.

[Ste01] M. Steyaert, B. De Muer, J. Janssens, M. Borremans, and P. Leroux, "The Design of CMOS Cellular Transceiver Front-Ends", in *Proceedings European Conference on Circuit Theory and Design*, Aug. 2001.

[Ste02a] M. Steyaert, Ph. Coppejans, W. De Cock, P. Leroux, and P. Vancorenland, "A Fully-Integrated GPS Receiver Front-End with 40 mW Power Consumption", in *Proceedings IEEE International Solid-State Circuits Conference*, pp. 396–397, San Francisco, USA, Feb. 2002.

[Ste02b] M. Steyaert, B. De Muer, P. Leroux, M. Borremans, and K. Mertens, "Low-Voltage, Low-Power CMOS-RF Transceiver Design", *IEEE Transactions on Microwave Theory and Techniques*, vol. 50, no. 1, pp. 281–287, Jan. 2002.

[Stu98] F. Stubbe, S. V. Kishore, C. Hull, and V. Della Torre, "A CMOS RF-Receiver Front-End for 1 GHz Applications", in *Proceedings of the IEEE Symposium on VLSI Circuits*, Honolulu, Hawaï, June 1998.

[Tin98] S. F. Tin, A. A. Osman, and K. Mayaram, "Comments on "A Small-Signal MOSFET Model for Radio Frequency IC Applications"", *IEEE Transactions On Computer-Aided Design of Integrated Circuits and Systems*, vol. 17, no. 4, pp. 372–374, Apr. 1998.

[Tsi87] Y. P. Tsividis, *Operation and Modeling of the MOS transistor*, McGraw-Hill, 1987.

[Van02a] P. Vancorenland, *MOSCAL V1.6 Tutorial*, 2002.

[Van02b] P. Vancorenland, P. Coppejans, and W. De Cock andM. Steyaert, "A Quadrature Direct Digital Downconverter", in *Proceedings Custom Integrated Circuits Conference*, pp. 235–238, Orlando, USA, May 2002.

[Van02c] P. Vancorenland, Ph. Coppejans, W. De Cock, P. Leroux, and M. Steyaert, "Optimization of a Fully Integrated Low Power CMOS GPS Receiver", in *Proceedings IEEE/ACM International Conference on Computer Aided Design*, pp. 305–308, San Jose, CA, USA, Nov. 2002.

[Van03] P. Vancorenland, *Low-Power Building Blocks for CMOS Wireless Receivers*, PhD thesis, K.U.Leuven, Belgium, May 2003.

[vdZ62] A. van der Ziel, "Thermal Noise in Field Effect Transistors ", *Proceedings of the IEEE*, pp. 1801–1812, Aug. 1962.

[vdZ86] A. van der Ziel, *Noise in Solid State Devices and Circuits*, Wiley Interscience, New York, 1986.

[Wan02] A. Wang, H. Feng, R. Zhan, G. Chen, and Q. Wu, "ESD Protection Design for RF Integrated Circuits: New Challenges", in *Proceedings Custom Integrated Circuits Conference*, pp. 411–418, Orlando, Florida, USA, May 2002.

[Wu98] S. Wu and B. Razavi, "A 900MHz/1.8GHz CMOS receiver for Dual-Band Applications", *IEEE Journal of Solid-State Circuits*, vol. 33, no. 12, pp. 2178–2185, Dec. 1998.

[Xu00] M. Xu, D. K. Su, D. K. Shaeffer, T. H. Lee, and B. A. Wooley, "Measuring and Modeling the Effects of Substrate Noise on the LNA for a CMOS GPS eceiver", in *Proceedings Custom Integrated Circuits Conference*, pp. 353–356, San Diego, USA, May 2000.

[Yue98] C. P. Yue and S. S. Wong, "On-Chip Spiral Inductors with Patterned Ground Shields for Si-Based RF IC's", *IEEE Journal of Solid-State Circuits*, vol. 33, no. 5, pp. 743–752, May 1998.

[Zho98a] J. Zhou and D. J. Allstot, "A Fully Integrated CMOS 900 MHz LNA utilizing Monolithic Transformers", in *Proceedings IEEE International Solid-State Circuits Conference*, pp. 132–133, San Francisco, USA, Feb. 1998.

[Zho98b] J. Zhou and D. J. Allstot, "Monolithic Transformers and Their Application in a Differential CMOS RF Low-Noise Amplifier", *IEEE Journal of Solid-State Circuits*, vol. 33, no. 12, pp. 2020–2027, Dec. 1998.